CONTENTS

INTRODUCTION

For some, the phrase, 'butterflies in my stomach' isn't just a metaphor. In Mexico, the larvae of the tequila giant skipper butterfly – known as white maguey worms – are a prized delicacy, rich in proteins, fats and essential minerals.

But most butterfly larvae are toxic and definitely shouldn't end up in your belly, particularly those that are brightly coloured or hairy. Any reference to butterflies-in-stomachs in this book is strictly figurative and represents one of the many puzzling biological phenomena that occur in our bodies.

Do you ever catch yourself lying awake at night, mulling over these things and other pressing questions of life – like 'Why do I stalk my ex on social media?' or 'When I talk to myself in my head, who's actually listening?' These late-night brain gymnastics about identity, relationships, and more, aren't explained solely by psychology or philosophy; they're deeply rooted in biology.

If you've never quite outgrown the 'why' stage and your sense of curiosity borders on the obsessive (picture Sherlock Holmes with a microscope), you're in good company here and exactly where you need to be.

Why Do I Get Butterflies In My Stomach? presents one of life's greatest mysteries, right up there with the Pyramids, and second only to why you thought that fourth cup of coffee was a good idea. Could this sensation be a profound, soul-deep connection with another human being? Or is it actually that discounted sushi you ate three hours ago?

As gloriously squishy, biological entities, anything that piques our curiosity about our bodies is obviously going to be biologically grounded – even if it's hiding behind a facade of mystery or myth. But there's a whole universe of little-known biological phenomena secretly driving the more obscure, which is what we're here to uncover.

Cue the concept of interoception, hailed by some as your sixth sense. First coined by neuroscientist Sir Charles Sherrington in 1906, it's your body's way of sensing what's going on inside you. It works with your nervous system, sending internal signals to the brain, which interprets and responds to them. Unlike the five senses that use specialised organs, it relies on neural pathways hidden deep within the body, helping you interpret feelings, mood and decisions based on signals from organs like the heart, lungs, gut, and even your face.

So, what's in it for you, dear reader? Hopefully, in addition to a newly discovered fondness for your own bodily quirks, you get a sample toolkit of your inner world, complete with logical self-assessment hacks, well-being tips and probing quizzes to indicate just how in tune you are with your own insides. For those craving change, there's a few habit busters and sensible transformational strategies, not to mention simple self-experiments and reflection exercises to really delve into your biology. These cheat codes help you not just to understand what's going on, but to do something about it.

Feel free to jump into whichever chapter tickles your fancy – there's no wrong way to start. Each question is unpacked with care and rigour to address the existential angst that first drew you there, and hopefully transform that restless flutter into a clearer, steadier perspective.

THE HEART
THE BIOLOGY OF ATTACHMENT

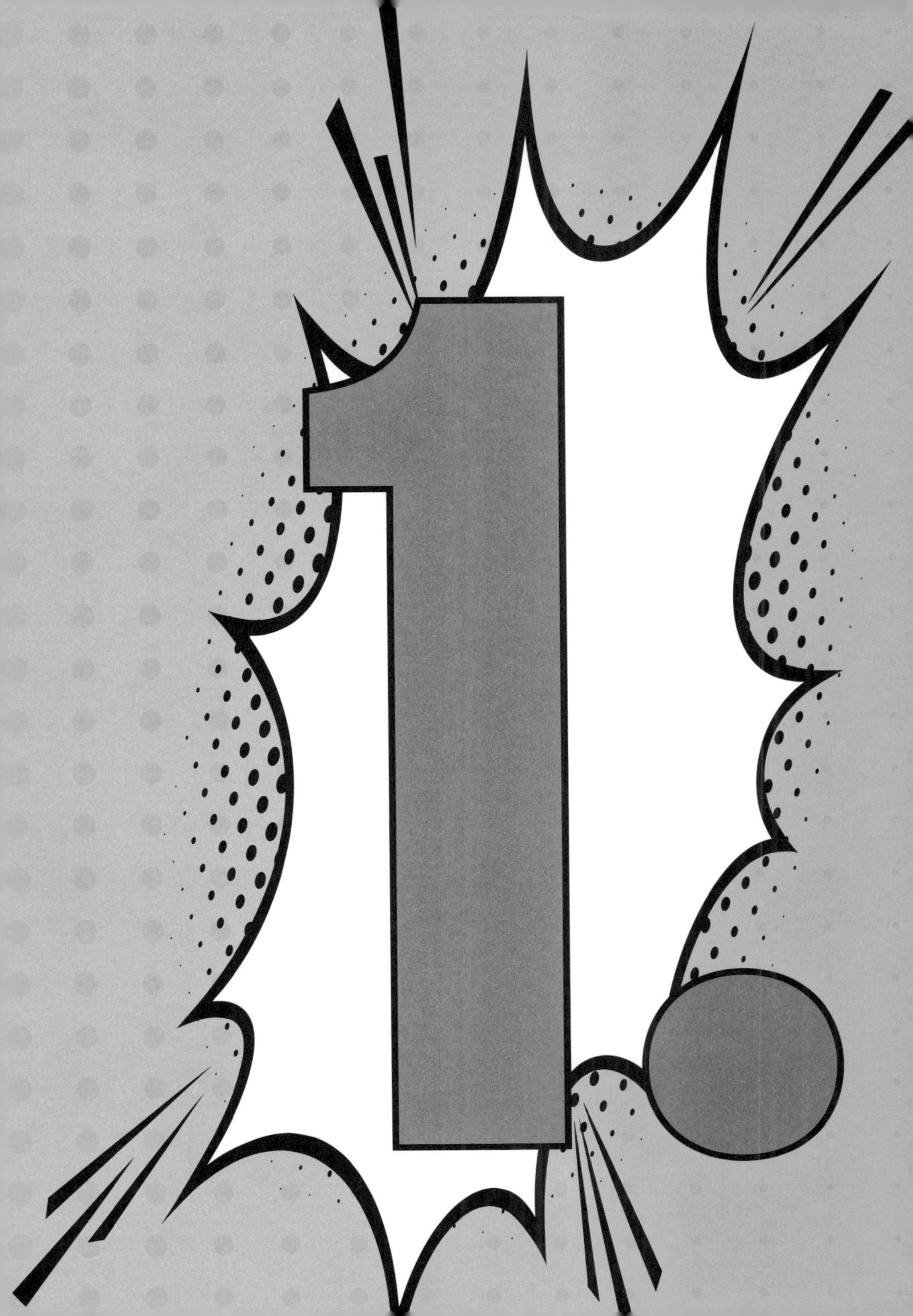
1.

WHAT IS LOVE, AND IS IT ALL IN MY HEAD?

So, you're about to pay for your morning coffee when you notice a piece of paper in your wallet or purse. It's a handwritten note that says:

You smile, thinking, 'Aw, they remembered!', and sigh silently, tapping in your payment as you drift into a transient state of bliss, comfort, connection and gratitude.

Clearly, when you love someone the mere thought of them can make you feel good inside. But what is actually going on inside your brain during this time? Is it playing tricks on you, or is something more profound at work here? For a state of being or emotion that is almost universally experienced, it's a question that is seldomly asked.

Love is not just of the romantic kind. According to ancient Greek society, at least eight different varieties exist; they encompass everything from self-love to the affection between couples, friends, family and strangers.

At its core, love is a survival mechanism orchestrated by the brain and body to promote behaviours that prioritize bonding, making babies and caregiving. Far more than just an ethereal feeling, love denotes a set of biochemical events that rely on an ancient cocktail of chemical substances in the brain, heart and body. Hormones, along with the brain's specialized chemical messengers (neurotransmitters and neuropeptides), engage with each other in an intricate dance to elicit this most complex and abundant of experiences.

BRAIN IMAGING STUDIES

You can't talk about the biological underpinnings of love without mentioning Dr Helen Fisher, the trailblazing American biological anthropologist who dedicated her life to understanding the brain chemistry of human connection and revolutionized how we understand it. She and her colleagues produced the first evidence of what goes on in the brain when you fall in love, dispelling the myth that love is some sort of supernatural event. Love, it seems, is something evolution has hardwired into your brain.

In 2005, Fisher led a team that published the first functional magnetic resonance (fMRI) images of the brains of individuals experiencing romantic love. By monitoring changes in the cerebral blood flow of participating college students, which indicated neural activity, she identified specific brain regions (such as the ventral tegmental area and the caudate nucleus, rich in dopamine) that became active after students viewed images of their beloved. This aligns with the theory that love isn't just an emotion but rather an evolutionary drive, to motivate partners to form and sustain deep bonds with their chosen mates.

Fisher's research laid much of the groundwork for the scientific study of love that continues to evolve, with ongoing research refining the roles of chemicals, such as oxytocin, vasopressin and serotonin, and brain circuits involved in love.

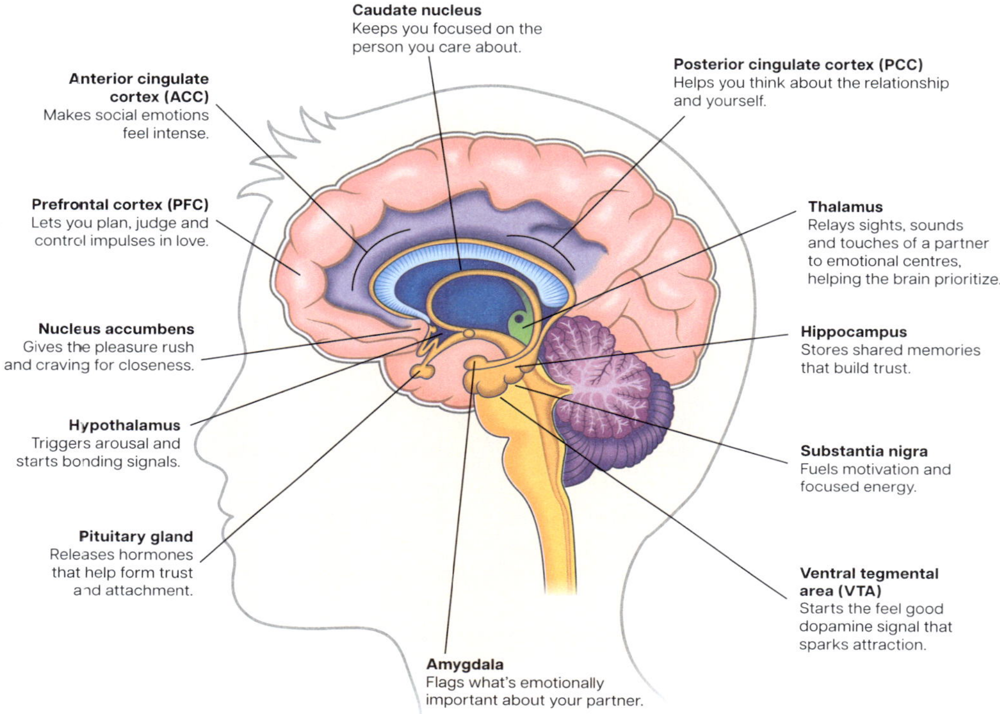

AREAS OF THE BRAIN MOST ASSOCIATED WITH FEELINGS OF LOVE
(insula and putamen not clearly visible in diagram)

While brain activity is central to understanding the neurobiology of love, it's not just a head thing – it's a full body experience. Hormones and other signalling compounds trigger powerful reactions that ripple through your senses and affect body systems, such as the cardiovascular and immune systems. But when it comes to love, the heart, in particular, plays a vital role.

YOUR SENTIENT HEART

As the universal symbol of love, the heart (long revered as the source of emotion, romance and spiritual connection) proves the ancients were right all along.

Nowadays, we know that the heart isn't just a passive recipient of brain signals – in fact, evidence shows that it engages in dynamic two-way communication with the brain and body, acting as a powerful amplifier of how we experience love.

Here's why this happens: your heart has its own built-in sensors, or special nerve cells, that detect changes in blood pressure (mechanosensory receptors) or blood chemistry (chemosensory receptors). Once activated, they send instant updates straight from your heart to your brain and spinal cord via a group of specialized nerve fibres called sensory neurites. Thousands of these fibres are spread throughout your heart and major blood vessels. As the heart's intrinsic nervous system, these fibres feed critical information back to the brain, influencing your ability to think and make decisions. It is believed that, through this feedback loop, cardiovascular arousal (such as a racing heartbeat) can amplify emotions, sending signals that intensify feelings such as fear or anxiety. For example, when you see your crush and your heart pounds, the brain interprets this as a threat, further turbocharging your nervous response.

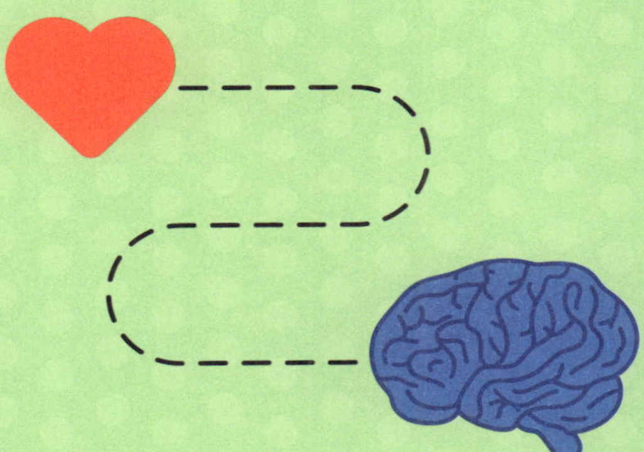

YOUR HEART'S MANY HATS

You'd think your heart already had enough things to do, but like a true overachieving multitasker it also moonlights as an endocrine gland, secreting a hormone called atrial natriuretic peptide (ANP). When the right upper chamber of your heart (atrium) stretches – from extra blood volume due to a spike in blood pressure, say – it releases ANP like a built-in pressure valve to restore balance.

During those warm and fuzzy moments of early love, the hormone oxytocin also nudges your heart into releasing ANP to help lower your blood pressure and induce relaxation. And recent research showing that your heart also secretes just as much oxytocin as your brain does further cement what the famous poets always knew: that love isn't just in your head; your heart's in on it too.

Greek philosophers like Aristotle and Galen viewed the heart as highly important, influencing emotions and even sexual love.

MISSION COMPLETE?

Or is it? According to Dr Fisher, lasting love doesn't happen by accident, it takes work. You need to constantly work with your brain chemistry, not against it. So, to keep the spark alive, prioritize regular intimacy (which drives up the oxytocin system), try new experiences together (the dopamine system) and maintain positive communication (which reduces stress hormone cortisol). Touch, novelty and kindness – three words that provide the science-backed recipe for long-lasting love.

THE NEUROBIOLOGY OF LOVE

Stage	Brain regions	Key chemicals	Body responses	Sensations
Initial attraction	Reward centres (NAc, VTA)	↑ Dopamine (pleasure) ↑ Noradrenaline (alertness) ↑ Adrenaline (physical arousal) ↓ Serotonin (obsession begins)	Racing heart, sweaty palms, adrenaline surge	'Butterflies', hyper-alertness
Mutual connection	Social processing (mPFC, insula)	↑ Oxytocin (trust/ bonding) ↓ Stress hormones	Calmer heart, synchronized breathing	Focused attention, warmth
Physical intimacy	Sensory areas (insula), reward centres ↓ Prefrontal cortex	↑ Testosterone (libido) ↑ Oxytocin ↑ Dopamine, endorphins (pleasure/bonding)	Rapid heartbeat, flushed skin Genital vasocongestion (swelling)	Arousal, tingling, euphoria, losing control
Deepening bond	Memory (hippocampus), reward centres	↑ Oxytocin ↑ Vasopressin (attachment)	Calmer heart rate, relaxed gut	Security, comfort, commitment
Long-term commitment	Decision-making (mPFC), reward centres	Stable oxytocin, endorphins	Better stress handling and possible heart benefits over time	Profound security, celebration
Parenthood	Reward/ maternal circuits (caudate nucleus, putamen, SN)	↑ Oxytocin, prolactin (bonding)	Heart synchronized with infant cries Uterine contractions (birth) Milk ejection	Overwhelming love, joy–exhaustion

mPFC (medial prefrontal cortex); NAc (nucleus accumbens); SN (substantia nigra); VTA (ventral tegmental area).

Arrows indicate activity changes: ↑ = increased activity; ↓ = decreased activity

A BRAIN (AND HEART'S) JOURNEY TO LOVE: FROM FIRST GLANCE TO PARENTHOOD

1 THE WEDDING GLANCE Your eyes connect over the pre-dinner champagne flutes at a friend's wedding. Your ventral tegmental area (VTA) and nucleus accumbens (NA) activate, sending dopamine to flood your reward circuit as you process their good looks.

2 SUCH A FLIRT You bump into them at the free bar. Your heart hammers against your ribs (thanks noradrenaline and adrenaline), palms sweat as adrenaline produced from your adrenal glands spikes. 'Butterflies' flutter in your gut because this stress response redirects blood from digestion to muscles.

You talk. Their laugh and flirting make your posterior temporal sulcus light up (social reward). It seems that they like you too. Oxytocin (produced by the hypothalamus and released into your blood by the pituitary gland) whispers 'trust this', silencing your amygdala's fear signals. Your hearts begin syncing rhythmically (see page 18). Pupils might dilate, and there may be flashed cheeks.

3 SLOW DANCE (SCENT & TOUCH) You can't quite believe you are both dancing together. Their hand on your waist sends signals through your insula (anterior insular cortex (AIC)) that integrates touch and scent with emotion. Vasopressin – also released by the pituitary gland – surges, tagging them as 'special'. During each pause of your heartbeat (diastole), your brain hyper-processes their woody fragrance. An electric warmth spreads through you as you sway to the beat in unison.

4 NUMBER EXCHANGE As digits flash on your screen, your hippocampus (involved in learning and memory) stamps this day firmly in its bank. In the taxi home your dopamine cravings spike because separation is quite the torment. Your heart suddenly skips a beat when you receive a text from them (at last!): that adrenaline does like to linger.

5 FIRST DATE Your amygdala awakens and is not for turning. Typical, today of all days your hair is misbehaving. 'Will they still think I'm cute?' Adrenaline frees some 'butterflies' when you see them at the venue. Your heart races as they catch you staring at them while they examine the drinks menu – its vibrations echoing in your insula, amplifying every millisecond of this awkward moment.

6 FIRST KISS Their lips meet yours. Your NAc detonates its dopamine dam, flooding you with pleasure. Oxytocin and vasopressin bonds you both mid-kiss. As your heart contracts, attention narrows to their tongue's flexibility. That soft lip pressure taps into deep neural pathways that link pillowy touch with safety and pleasure. As your prefrontal cortex tunes out, time itself momentarily dissolves. With serotonin thrown in everything feels wonderful.

7 OBSESSED MUCH? They're all you can think about. Serotonin levels plummet to OCD-like lows. They love the bright lights of the city; you love the countryside. They're not a fan of little ones; you can't wait to have your own – it'll never work. Dopamine surges force the more cautious/critical parts of your prefrontal cortex offline, helping you ignore this and other amber/red flags. You're beginning to display the hallmarks of stress – the noradrenaline in your system makes you forget meals, sleep less and your heart pound at every phone call/notification, further feeding the frenzy. In bed, systole (heart contraction) sharpens mental imagery, and you fantasize about them in a very intimate way.

8 GETTING IT ON Testosterone and oestrogen finally move the goalpost from third base to home run. Your hypothalamus orchestrates the rush of oxytocin that seeps from your pituitary gland (your heart secretes it too). Racing hearts begin to sync and you mirror each other's skin conductance. It ends like this: a burst of endorphins, serotonin, dopamine and oxytocin totally infuses, draping you in euphoric peace.

9 BECOMING EXCLUSIVE Your ventrolateral prefrontal cortex (vlPFC) – the part of your brain that helps you suppress inappropriate responses and keeps you focused – takes one look at that sexy stranger and goes 'Nope. Not today!' Vasopressin locks in mate-guarding – your eyes are firmly on one prize. Your heart rhythm stabilizes as cortisol drops. A deep calm settles, signalling security is near.

10 MEETING THE FAMILY Your temporoparietal junction (which integrates sensory signals for social awareness) decodes in-laws' expressions. Adrenaline briefly spikes, causing your heart to pound during introductions. Oxytocin soothes as you feel accepted, and your heart steadies as approval materializes.

11 LIVING TOGETHER Your anterior cingulate cortex (ACC) helps you to not ask for a cup of tea when you spot them frantically tidying – it decodes stress signals like a social survival expert. It also pieces together why they cranked the heating way down: 'Was it the energy bill or my passive aggressive sigh when they turned it up last week?' Your hippocampus stamps the scent of their shampoo on your sheets as 'home'.

12 MARRIAGE Hearing their vows reactivates your NAc/VTA leaving you with that same dizzy rush you felt when you first locked eyes at your friend's wedding reception. Endorphins gild the day in gold, and as the sky lanterns rise, marking the end of the day's celebration and the dawn of a new chapter, your heart will over time reap the rewards of this union: a potential shield against disease.

13 AND THEN THERE WERE THREE Now it's gone full circle. The bond that first connected you both together shares deep roots with that primal love between mother and child (these early parent–infant neural circuits laid the foundation for adult pair-bonding).

Maternal neural circuits eclipse romantic love. Mum's primary focus is on baby. Oxytocin and prolactin help make nursing possible.

Your caudate nucleus (which encodes and reinforces positive social rewards) lights up when baby smiles and giggles. Your putamen (coordination and fine motor control) primes you to catch them mid-fall. With a suppressed amygdala, dopamine from your substantia nigra (reward processing) fuels an inexhaustible drive to keep your care and love going – no matter the cost.

CAN HEART RHYTHMS SYNC BETWEEN PEOPLE?

The cardiac pacemaker cells of your heart synchronize each heartbeat, because coordinating contractions maximizes blood flow with minimal energy expenditure. When this synchrony, referred to as 'cardiac synchrony', breaks down during an arrhythmia, say, the heart stumbles into chaotic rhythms that waste energy and increase the chance of it malfunctioning. But your heart also cares about your existence on a much grander scale, syncing its rhythms with that of other people – so that they beat at the same time and speed.

A 2021 study led by Dr Jacobo Sitt sought to find out whether people's heart rates sync when they listen to or watch the same story (narrative stimulus). They found that when people paid close attention to excerpts from the audiobook *20,000 Leagues Under the Sea*, their heart-rate patterns showed significantly similar fluctuations, which didn't occur as much in those who were deliberately distracted. This synchrony was also not due to shared breathing patterns, as their respiration rates did not synchronize in the same way. They also found that when a person's heart rate matched more closely with that of others, they tended to remember the story much better.

TRY THIS:

DOES YOUR HEARTBEAT SYNC WITH OTHERS?

Consider doing this task only if you have no history of heart issues. If in doubt, skip:

Step 1. Finding your pulse. Turn your left palm up. Place your right index and middle fingers lightly on the thumb side of your wrist (just below the heel of your hand). Feel the soft thump-thump of your radial artery.

Step 2. Measure your resting heart rate. Set a timer for 60s and count every thump you feel. Record your number. This is your resting heart rate in beats per minute (bpm).

Step 3. Listening to fast/slow heartbeats. Search on YouTube for fast heartbeat (150bpm) or slow heartbeat (40bpm) sounds on hospital demos or verified anatomy channels. Using headphones, play the fast heartbeat sound for 1 minute, measure your pulse and record your bpm. Do the same for 3 and 5 minutes. Rest for 10 minutes or until your pulse normalizes. Repeat, this time using the slow heartbeat sound.

Chances are your heartbeat either sped up or slowed down to mirror the heart rate you listened to. This phenomenon, known as 'physiological or heart rate synchrony', is something your heart is very good at doing. But why?

This intrinsic need to sync with others appears to be a natural byproduct of social interaction: an inbuilt social survival toolkit designed to foster close relations and promote social cohesion.

Heartbeat sound	Resting bpm	1min bpm	3min bpm	5min bpm
Fast				
Slow				

The small Sitt study of 94 participants suggests that, when you're really tuned in to a story or movie, not only is your brain engaged but your heart begins to beat in a similar pattern to others who are listening. The more attentive and engaged you are, the more your heart rate syncs with that of your peers, and this synchronization also predicts how well you will remember the story later.

Sharing an experience with someone can make your hearts sync up in this way. Take singing in a choir, for example, which requires that you hit the right notes together. A song's structure calls for coordinated physical effort and timing that will naturally synchronize breathing and heart rhythms.

Regardless of what activity you end up doing with others, the strength of any heart synchrony (how closely matched the heart rhythms are) appears to depend on the quality of the relationship between the people doing the syncing. For instance, a 2022 study published in *Nature Human Behaviour* revealed a fascinating link between heart rate synchrony and romantic attraction during real-life blind dates. The research led by Professor Mariska Kret equipped participants with eye-tracking glasses and wearable devices to measure heart rate and skin conductance. The pairs engaged in conversation and silent gaze-holding while the sensors tracked their physiological responses. Surprisingly, the strongest predictor of mutual attraction wasn't overt cues such as smiling or flirting, but the synchrony of their heartbeats and skin conductance responses, both of which are governed by the autonomic nervous system (ANS).

Your ANS regulates your heart rate via two branches: the sympathetic branch, also known as fight or flight, which speeds your heartbeat up during excitement, and the parasympathetic branch, also known as rest and digest, which slows it down during relaxation.

Research using windowed cross-correlation analysis (a technique that measures real-time alignment between physiological signals such as heart rate, skin conductance and pupil size) has shown that when you socialize, your ANS and that of others synchronize and go through parallel shifts together. Shared excitement causes people's bodies to fire up in sync, while mutual relaxation shifts their systems collectively into a shared 'rest-and-digest' state.

Mirror Neuron

A type of brain cell that fires both when an individual performs an action and when they observe another performing the same action.

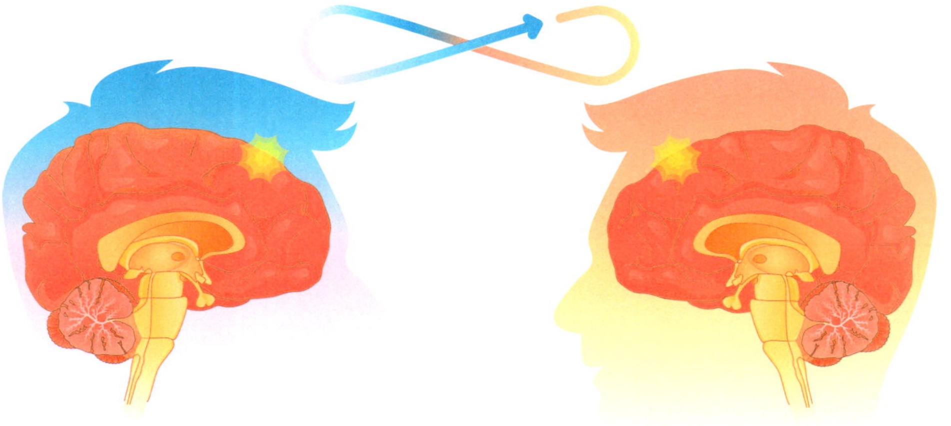

NEURAL MIRRORING

Neural mirroring is when people connect emotionally or empathize with each other, and neurons in their brains become active and cause subconscious mimicry, not just of facial expressions or gestures but also indirectly of internal physiological states such as heart rhythm. For example, an infant's emotional cues trigger the mother's brain to 'mirror' their state by firing similar neurons in her brain, which help her to understand her baby's actions and intentions, which in turn helps trigger similar feelings and actions, thereby causing the ANS to respond in much the same way.

ASYNCHRONOUS HEART

Do you feel socially disconnected from certain people who you know well? If so, it's likely that your heart rhythms do not sync. There are situations when people just do not 'vibe' with each other and physiological synchrony fails to occur. The reasons for this vary, but having issues with socializing might not be the only reason for this asynchrony.

The Sitt study (see page 18) also explored more practical applications of heart synchrony among patients with disorders of consciousness: for example, those in a coma or vegetative state. These 19 patients showed much lower levels of synchrony when listening to a story compared to 24 healthy control individuals, which suggests conscious processing of narrative stimuli is required for synchrony to occur. This finding could be developed into a simple way of finding out if someone is consciously processing information, which could in turn provide a valuable clinical tool to diagnose levels of consciousness in patients who aren't able to communicate in usual ways.

Reasons for absent or reduced heart synchrony	
Social disorders	People with social anxiety disorder or with autistic traits often show reduced or reversed synchrony during social or close interactions. They find it difficult to 'tune in' with others, which might be reflective of their impaired emotional/cognitive ability to engage.
Relationship woes	Romantic partners exhibiting low heart rate synchrony often report lower relationship satisfaction and empathy. In couples therapy, this could serve as an objective biomarker of relationship distress.
Distraction	Two people watching the same film with one distracted by their phone won't sync because of lack of shared focus.
No emotional bond	Strangers collaborating on a task sync less than close friends due to absence of trust and empathy that comes with emotional closeness, which is known to strengthen physiological synchrony.
Physical distance	A highly speculative idea is that online meetings lack the electromagnetic cues that present in-person thereby prohibiting signal exchange.

THE SIGNALS YOU'RE SENDING WITHOUT EVEN KNOWING IT

During lunch with friends, you notice one person's torso is angled towards another person. When you speak, they only turn their head, never their body. Even after you ask a question, their response still seems aimed at the other friend. Without realizing it, they are sending you a signal, and the message is: 'My focus, comfort level and loyalty lie with the other person, not you.'

Your heart's a bit like that. Not in the sense of closing itself off to you (that wouldn't be a good idea), but in the way that it's able to send messages to your brain and body that influence how others perceive and respond to you without you even knowing. Right now, as you're reading these words, your heart is sending thousands of messages to your brain. Not just the steady 'padam-padam' you might notice if you pause to listen, but a sophisticated stream of data that's quietly shaping every decision you make, every emotion you feel and every relationship you navigate.

So, why on earth is your heart doing this, and how? Well, your heart is a loquacious beast of an organ, in a nonstop exchange with your brain and body. This constant two-way communication affair occurs through multiple pathways. Without these connections, your heart wouldn't have any autonomy over itself or be able to influence how you feel or think, and this is all thanks to it having its very own 'little brain'.

THE ICNS

Nestled deep within your chest, dotted around chambers and valves typically associated with circulation, lies the intrinsic cardiac nervous system (ICNS) or 'brain inside the heart'. It's a fitting name when you consider what it contains – complex neuronal structures filled with different types of nerve cells and fibres (about 40,000 in total) organized into interconnected clusters called 'ganglionated plexi'. Housed within fat pads, they lie on the muscle surface of the heart and operate as independent information processing centres that play a major role in controlling your heart rate and rhythm.

What processes information, makes decisions and executes action based on sensory data?

Not your brain – your heart. Its ICNS operates like a mini-brain complete with its own primitive intelligence system, making split-second adjustments with every single beat.

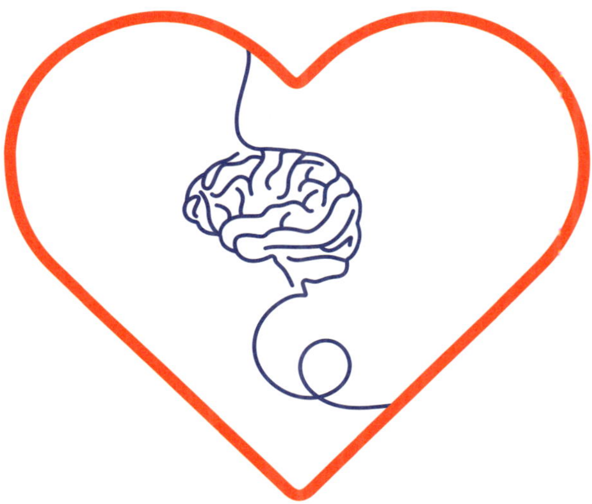

There are three types of nerve cells (neurons) and their fibres:

1. THE SCOUT (SENSORY NEURONS)

JOB: Heart's vigilant lookout

SPECIALITY: Monitors stretch, pressure and blood chemistry with every heartbeat

COMMUNICATION: Sends sensory data up to your brain via the vagus nerve and spinal cord

PERSONALITY: Never sleeps, always reporting

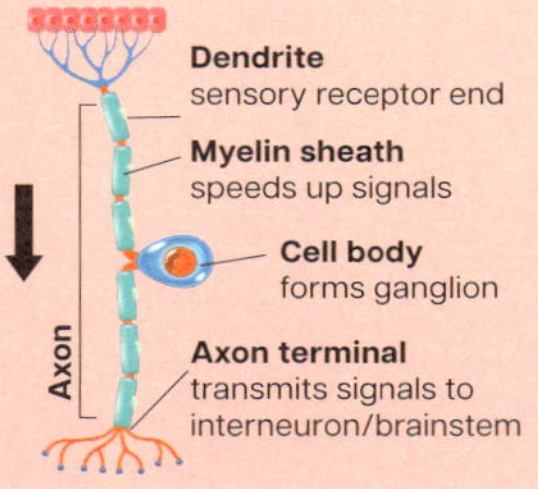

2. THE EDITOR (INTERNEURONS)

JOB: Local information processor

SPECIALITY: Amplifies urgent signals, quietens the noise and combines multiple inputs

SUPERPOWER: Shapes how your brain interprets your heart's activity

AMAZING FACT: Exists only in your heart, brain and spinal cord – nowhere else!

SECRET SKILL: Triggers reflexes (like blood-flow adjustments) without asking your brain's permission, and can bypass the brain and directly trigger ANS motor neurons

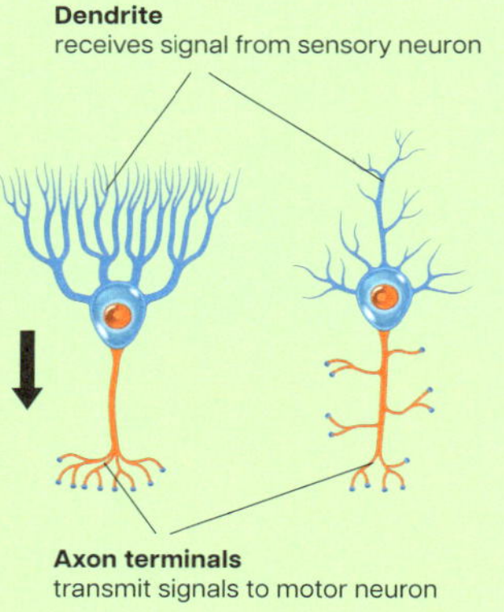

3. THE RESPONDER (MOTOR NEURONS)

JOB: Your brain's remote-control system

MODES: Sympathetic: speed up – fight-or-flight surge

Parasympathetic: slow down – rest-and-digest mode.

AMAZING FACT: Works 24/7 in the background with no conscious effort required

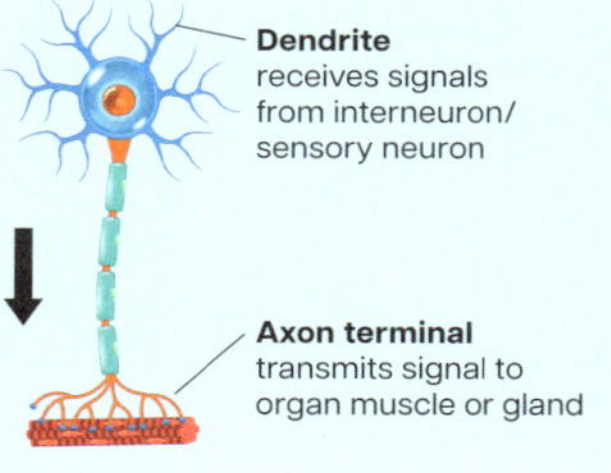

Next time your heart races before a presentation, remember: your Scout detected the changes in your blood pressure, your Editor decided it was important and your Responder hit the accelerator – all before your conscious mind caught up.

HRV

Your heart might seem to beat like a metronome, but it's a dynamic, responsive organ. Each beat varies slightly in timing – this is heart rate variability (HRV). These subtle shifts reflect the ongoing dialogue between your brain and ANS, showcasing just how flexible and responsive your heart truly is.

SIGNALS YOU'RE BROADCASTING

Nonverbal cues, such as posture, voice tone and eye contact, reveal your inner state. But your physiology does too. Your heart rate, skin temperature, moisture and blood oxygen levels often betray what your words try to hide.

Remember the last time you told someone you were 'fine' when you clearly weren't. Your conscious mind might have crafted that response, but your heart was telling a completely different story. Unbeknown to you, its rhythm was somewhat erratic, signalling a degree of internal chaos to anyone paying attention – including your own brain, which was trying to make sense of the mixed messages.

The disconnect between what you say and what your heart is signalling creates what some scientists call 'incoherent' HRV. When your heart rhythm is chaotic and unpredictable, it sends stress signals racing up your vagus nerve to your brain's emotional centres. The result? Impaired thinking, reduced creativity and that nagging sense that something's just not right.

By contrast, 'coherent' HRV – smooth, steady fluctuations – means your body is better equipped to manage stress. It supports clearer thinking, unlocks creativity and keeps you emotionally strong.

YOUR HEART'S DIRECT LINE TO DECISION MAKING

Here's the kicker: your heart helps shape decisions, often before your conscious mind even has a chance to weigh in. Each heartbeat generates tiny electrical signals in your brain, perfectly timed to its rhythm, known as heartbeat-evoked potentials (HEPs). These signals act as a vital communication link between your heart and brain, and they are shaped by your emotions and breathing.

When you're facing a tough decision, your heart is already providing your brain with essential internal data, helping to guide your choices. It doesn't just react to what you feel but also actively helps shape those emotions. By adapting its function to your most recent experiences and the chemical environment in your body, your heart integrates its short-term memory, HEPs, sensory inputs and even hormones (such as oxytocin, which fosters connection, and ANP, which reduces stress) and sends its recommendations through multiple pathways to your brain's decision-making centres.

But what about your emotions? Well, your heart watches your HRV to gauge whether you're relaxed and coherent, or stressed and scattered, and then feeds a real-time report into key emotion-regulating centres in your brain (the Central Autonomic Network, including areas such as the insula and prefrontal cortex). Together they dial your emotional state up or down.

TUNE IN

The secret to tuning into your heart's wisdom lies in understanding something called heart coherence. When your heart rhythm (as seen on an ECG) becomes smooth and wave-like rather than erratic and chaotic, it signals to your brain that all systems are functioning optimally. This coherent state, like a high HRV, enhances mental clarity, boosts creativity and improves your ability to regulate emotions.

RELATIONSHIPS

Your heart's influence extends far beyond just your individual experience; because you're a social beast, it's perfectly attuned to the social and relational environment around you. When you're with people who make you feel safe and valued, your heart responds by shifting into coherent patterns that enhance your capacity for connection and empathy.

Brain imaging studies reveal something remarkable: when you're with trusted attachment figures, the pain and distress centres in your brain quieten down while the safety and connection centres light up. Your HRV improves, your oxytocin production increases, and your entire nervous system shifts into a state that promotes healing and growth.

Conversely, when relationships are strained or threatening, your heart detects the danger and responds accordingly. The chronic stress of difficult relationships doesn't only feel bad but creates measurable changes in your heart's rhythm patterns that can persist long after the immediate interaction ends.

TRY THIS:

You can actually train yourself to achieve heart coherence through surprisingly simple techniques. Try this right now: place your attention on your chest area and begin breathing in a steady rhythm – perhaps five seconds in, five seconds out. As you maintain this pattern, deliberately cultivate a feeling of gratitude or appreciation. Notice how this shifts not just your breathing, but your entire sense of well-being.

Start small. Several times throughout your day, pause and ask yourself: 'What is my heart telling me right now?' Notice the quality of your heartbeat, the rhythm of your breathing, the overall sense of expansion or contraction in your chest. These physical sensations provide information about your emotional state and your body's assessment of your current situation.

Remember, this isn't just a relaxation/heart assessment exercise, but rather an opportunity to retrain the communication pathways between your heart and brain. With practice, you can learn to recognize the subtle difference between your heart's coherent and incoherent states, giving you real-time feedback about your emotional and physiological condition.

Category	Positive relationship signs (safe and connected)	Negative relationship signs (threat and stress)	Awareness and action steps
Heart rhythm	Smooth, steady beats; high HRV	Erratic 'flutters'; low HRV	• Track HRV daily via wearable device (e.g. smartwatch) • Practice 5–6 breaths/min breathing • Do heartbeat meditation (heart-focused breathing)
Resting heart rate (in a young, fit, healthy person)[1]	Lower (55–65 bpm); calm presence	Elevated (75–85+ bpm); on-edge feeling	• Monitor resting rate trends • Set smartwatch stress alerts
Stress response	Quick recovery; smaller spikes to triggers	Slow recovery; exaggerated reactions	• Note stress-trigger patterns • Use 30s breathing pauses at first sign of stress
Body–breath connection	Heart syncs with breath: speeds on inhale / slows on exhale (inhale ↑/ exhale ↓)	Shallow breathing; no heart–breath sync	• Check natural breathing rhythm with partner • Do co-regulation breaks (sync breathing)
Partner synchrony	Heart rates align; 'same wavelength' feeling	Mismatched rhythms; lonely together	• Practise joint biofeedback (joint breathing exercises while monitoring HRV) • Create positive rituals (e.g, gratitude check-ins; 5s handshake)
Physical sensations	Steady, grounded feeling; warm tone of voice	Chest jolts/heart flutters; tight voice/ posture	• Log 'flutter' triggers immediately • Notice changes in body tension before emotion
Neuro-chemical state	Oxytocin-driven calm; effortless connection	Adrenaline/cortisol surges; lingering agitation	• Acknowledge 'safe default' moments • Rewire habits via shared laughter/positive touch

[1] A normal resting HR lies between 60-100 bpm and varies from person to person depending on their age, fitness and health.

WHY IT'S NOT GOOD TO BE LONELY

It's a Saturday evening and you're kicking yourself for agreeing to accompany a friend to a private function. A night spent on the sofa, with a carton of ice cream and a good movie, sounds like pure heaven, while the reality of having to endure an evening of small talk, prying questions and endless chit-chat with strangers fills you with absolute dread. One of these two scenarios is more likely to kill you than the other – and it isn't the one with the ice cream.

Being alone is not the same as feeling lonely. If you choose solitude, your desired level of social connection matches your reality. No complaints there. But for many people, especially older adults, there is often no choice in the matter. And you don't have to be physically separated from people to feel lonely, either. Even if you're surrounded by people or have an amazing network of friends, you can still feel profoundly lonely if those relationships lack the depth or intimacy you truly need. The impact of social anxiety on your health can be just as devasting as being lonely.

Loneliness is a big problem. It affects millions of people worldwide and is particularly worrisome if you feel lonely 'often or always', because that makes it chronic and turns it into a very different kind of beast. One that could potentially make your chances of dying within the next 10 years about 15 to 50 per cent higher than those who don't feel lonely all the time. In terms of increasing heart-disease risk, research suggests that it is comparable to smoking or being obese. And that's not all. If you're a cancer survivor, severe loneliness makes your chance of dying about 67 per cent higher than that of survivors who don't feel isolated.

The changes caused by loneliness don't always take decades to show up. Some of these effects can be seen within just a few years, meaning loneliness can have a relatively fast impact on your body's biological ageing.

As social animals, we are hardwired to thrive in a community and connect with others. Socializing matters because it makes you feel safe and switches on your parasympathetic nervous system (PNS), which induces calm and reduces stress (see table below). When this doesn't happen over a prolonged period, it triggers a set of complex biological pathways that ultimately lead to inflammation, which plays a big role in sending your general health south.

Body part	What the PNS does	Why it makes you feel calm
Heart	Slows your heartbeat	A slow heart tells your body you're safe
Lungs	Slows your breathing	Slow breaths naturally relax the body
Stomach and gut	Starts digestion again	Your body digests only when it feels safe
Eyes	Makes pupils smaller	Less light and less alertness = calmer feeling
Blood vessels	Lowers blood pressure	Reduces the 'on-edge' feeling in the body
Liver	Slows glucose release; shifts towards storing energy	Body moves out of 'fuel-burning' stress mode and into 'rest and rebuild' mode
Uterus	Relaxes uterine muscles	Reduced cramping/tension supports an overall sense of ease and safety
Urinary system (bladder and sphincters)	Bladder muscles relax; sphincters loosen to allow urination	Urination happens only when the body feels safe, so this reinforces the 'no threat' state

HOW DO YOU SOLVE A PROBLEM LIKE INFLAMMATION?

On paper, inflammation is meant to help – it's a natural part of your body's healing process. When you're sick or injured, your immune system sends special cells to repair things, which is why you might get a fever: inflammation is on the scene. But left unregulated for a long time, it becomes disruptive, turning a once protective process into a persistent background noise that wears down systems over time.

Your body does try to solve things before that happens by recruiting a trio of glands known as the HPA axis that link parts of your brain (hypothalamus and pituitary glands) with your adrenal glands sitting on top of the kidneys. Your adrenals release hormones such as cortisol into the bloodstream, which in short bursts can suppress the production of pro-inflammatory molecules called cytokines, such as IL-6 and TNF-α, to help you manage stress and inflammation. But like all overwhelmed systems, inflammation becomes an agitated non-conformist and begins to damage healthy tissues, ultimately contributing to long-term diseases, particularly those of the heart. Furthermore, your immune system behaves in odd ways, with some parts becoming overactive (causing inflammation) and other parts becoming weakened to the point that it reduces your ability to fight off viruses, making you more vulnerable to infections.

Chronic loneliness is one of those unfortunate states that is so detrimental it can impact your well-being in a major way. The problem isn't just that you feel sad, it's that it puts your entire body into a harmful threat mode. As a 'psychosocial stressor', loneliness keeps your sympathetic nervous system firing for far too long, which leads to a surge in stress hormones such as adrenaline, which boosts your heart rate and blood pressure to prep you for fight or flight, and noradrenaline, which tightens blood vessels to maintain blood pressure and help you stay alert.

It should come as no surprise, therefore, that the chronic stress and inflammation caused by loneliness can have profound cardiovascular consequences. These effects include heightened risks of hypertension, heart disease and diabetes. The continuous strain on the body can trigger a cascade of disruptions in psychological functioning, the neuroendocrine system (the network of brain–hormone signals that regulates stress and other body processes), the autonomic nervous system (which automatically controls vital functions), as well as metabolic and immune responses – all of which directly contribute to the development and worsening of these life-threatening conditions.

Loneliness → Stress → Inflammation → Disease

Body system affected	What chronic loneliness does	How this links to hypertension, heart disease and diabetes
Overall stress response	*Keeps fight or flight constantly active:* • Overactivates the stress response (sympathetic nervous system – SNS) • Weakens the calming response (parasympathetic nervous system – PNS) • HPA axis chronically activated, disrupts the natural cortisol rhythm (normally peaks upon waking and drops throughout the day)	*Sustained stress fuels damaging inflammation:* • Constant cortisol fails to control inflammation • Overactive SNS produces pro-inflammatory molecules • The vagus nerve no longer stops the release of cytokines when the PNS is weakened
Immune system	*Triggers chronic low-grade inflammation:* • Increases inflammatory markers (CRP, IL-6) • Causes 'conserved transcriptional response to adversity' (CTRA): immune cells shift what they produce to favour inflammation and reduce their virus-fighting ability	*Chronic inflammation damages the body:* • High markers indicate ongoing vessel damage • Weakens anti-viral defences
Mental health	*Creates hypervigilance and distress:* • Brain constantly scans for threats • Strongly linked to depression and anxiety	*Mental distress worsens physical stress:* • Depression increases inflammation, clotting risk and cortisol, raising heart disease risk
Ageing	*Accelerates cellular ageing:* • Driven by chronic inflammation	*Increases vulnerability:* • Makes the body more prone to age-related diseases like heart disease
Metabolism	*Disrupts blood sugar and fat storage:* • Raises long-term blood sugar (HbA1c) • Can increase body fat	*Inflammation causes insulin resistance:* • Cells stop responding well to insulin • Can lead to type 2 diabetes over time
Cardio-vascular	*Directly strains heart and vessels:* • Raises blood pressure • Increases blood clotting proteins • Lowers heart rate variability (stress resilience) • Causes oxidative stress (cellular damage)	*Inflammation and stress cause damage:* • Damages vessel linings, promoting plaque • Leads to hypertension, clots, heart attack and stroke

THE WOMAN WHO DIED FROM A BROKEN HEART

Even though it might sound like the stuff of tragic fiction, it's entirely possible for someone to die from a broken heart. Rare? Yes. But the condition – broken heart syndrome, or takotsubo cardiomyopathy (TC) – causes sudden short-term failure of the heart muscle in an otherwise healthy individual.

TC feels like a heart attack, with chest pain and breathlessness, but unlike a true heart attack, the arteries aren't blocked. The one distinguishing factor that shows up on imaging (echocardiogram or MRI) is the appearance of your left ventricle when it contracts. Instead of looking like a V-shape, it balloons at its tip and ends up with a circular bottom and narrow neck – much like a Japanese octopus trapping pot, or *takotsubo*.

First described in Japan in the early 1990s, TC accounts for 1 to 2.5 per cent of suspected heart attacks. Postmenopausal women make up 80 to 90 per cent of cases, and 96 per cent of those recover in weeks. But for the unfortunate few, it can persist and lead to complications such as:

- Heart failure
- Abnormal heart rhythms in the ventricles (ventricular arrhythmias)
- Inability to effectively pump blood throughout the body (cardiogenic shock)
- Blood clots in the heart (apical thrombus formation)

Rarely, TC can lead to cardiac rupture. This complication disproportionately affects elderly women and is more often than not fatal. A 2024 review by University of Virginia researchers uncovered 44 recorded cases of cardiac rupture in TC patients – 64 per cent of them died.

STRESS TRIGGERS

When TC was first discussed in a Japanese scientific review in 1991, it was not associated with any deaths. Rather it was considered a somewhat benign condition, linked to stress events, exhibiting symptoms similar to a heart attack.

Stress and emotional trauma have long been known to damage your cardiovascular system and cause heart problems. Over time they can flood your body with hormones that damage blood vessels, raise blood pressure and increase inflammation, contributing to plaque buildup in your arteries. They also throw your autonomic nervous system (ANS) out of balance, which can mess with your heart's rhythm and put constant strain on your cardiovascular health. In contrast, the emotional or physical stressors that trigger TC are often much more sudden.

Interestingly, the Covid-19 pandemic saw a dramatic rise in TC cases, which researchers think may be due to the combined effect of increased physical and emotional stressors at this time amplifying its occurrence.

Stressor	Trigger examples
Emotional	Death of a loved one, relationship conflicts, receiving distressing news
Physical	Severe illness, surgery, medical procedures
Neurological	Brain haemorrhage, stroke
Substance related	Heroin, amphetamines withdrawal
Infectious disease	Pneumonia, meningitis

WHAT HAPPENS IN YOUR BODY TO SUDDENLY DAMAGE YOUR LEFT VENTRICLE? AND WHY THE LEFT ONE?

Your left ventricle is rich in beta-2 adrenergic receptors, which are sensitive to adrenaline and other stress hormones. The shock of a loved one's death can trigger your brain's stress centres (HPA axis) to kick the sympathetic nervous system into overdrive, flooding your blood with stress hormones (adrenaline and noradrenaline). At very high levels, these hormones do some very bad things to your heart:

1. **'Catecholamine-induced cardiotoxicity'**: stress hormones bind to your heart's beta-receptors, which forces its muscle cells (cardiomyocytes) to overload themselves with calcium. This damages their energy factories (mitochondrial injury), which leads to the formation of highly reactive toxic molecules (reactive oxygen species) that damage the cells and effectively stun the heart muscle into a kind of temporary paralysis.

2. **'Microcirculatory dysfunction and coronary vasospasm'**: the hormone surge causes tiny blood vessels in the heart to spasm and malfunction. This stalls critical blood flow in your heart, which ultimately stuns the heart tissue and causes a mismatch in the supply of oxygen versus demand.

This devastating one-two punch to your heart, which first poisons the heart muscle itself and then causes the tiniest blood vessels in your heart to collapse, forces your heart into the characteristic balloon-like shape that defines TC – from which the more technical term 'apical ballooning syndrome' derives. Unlike a classic heart attack, where a single blocked artery creates a predictable pattern of damage, TC jumps across multiple regions, because the catecholamine surge hits your entire heart muscle and its microscopic blood vessels simultaneously.

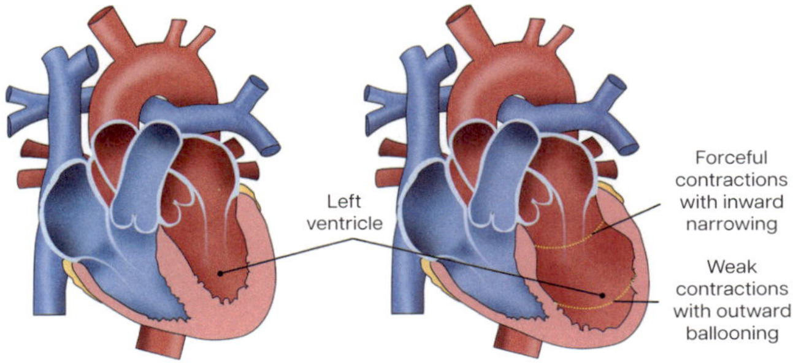

NORMAL HEART

The heart squeezes evenly from top to bottom during systole

APICAL BALLOONING SYNDROME
(Broken Heart Syndrome)

This shape happens when the heart contracts during systole

WHY SINGLE OUT POSTMENOPAUSAL WOMEN?

TC most likely targets this cohort because at this age oestrogen levels fall in women. Normally, oestrogen helps switch on genes that produce proteins that shield your heart cells from damage by stress hormones, calcium overload and harmful free radicals. Once oestrogen levels drop, that genetic 'bodyguard' system weakens, making the heart more vulnerable to sudden surges of stress.

When men do develop TC, it appears that their outcomes are far worse. A US study of nearly 200,000 hospitalized patients with TC found 6.5 per cent died in hospital, with men facing double the risk: their mortality rate hit 11.2 per cent compared to just 5.5 per cent for women. While these findings spotlight a potentially troubling gender gap, it doesn't necessarily follow that this applies to every TC case, because these percentages represent a group average from hospitalized patients and should not be used to predict an individual's outcome. An individual's prognosis depends far more on their specific health profile, the timeliness of care and the particular form of the condition than on their sex alone.

WHY DO SOME PEOPLE DEVELOP TC UNDER STRESS AND OTHERS DON'T?

It takes a 'perfect storm' of vulnerabilities triggered by intense stress for TC to occur. A sudden surge of stress hormones (catecholamines) can overwhelm your heart if you have underlying weaknesses, including:

1. Pre-existing flaws in the small vessels of your heart (they don't dilate properly)

2. Heart muscle cells prone to catecholamine toxicity (latent biological trait)

3. Heightened sensitivity to stress hormones surges, due to:

- Low oestrogen levels (reduced natural cardioprotection)
- Inherited genetic traits that affect your heart's function/ stress response (such as variations in genes involved in catecholamine signalling/clearance and maintaining your heart's structural integrity)
- Chronic neurological/inflammatory conditions (your heart is already primed to hyper-respond)

4. Overactive stress centres in the brain tend to aggressively stimulate your heart's nerves

5. Low psychological resilience (low stress tolerance/coping mechanisms)

SO, WHAT CAN YOU DO TO SAFEGUARD YOURSELF?

It seems pretty difficult to know whether you have this set of vulnerabilities, so how can you protect yourself? While medications exist to treat TC, your most powerful shield against developing it is stress management, which has proven cardiovascular benefits. Clearly anything that helps to keep your stress levels in check would indirectly help lower your chances of developing abnormal stress responses or any severe complications from TC. Engaging in activities that turn on your parasympathetic nervous system seems to be key. Most of the hacks highlighted below show promise as general supportive stress control and cardioprotective tools. But, in all honesty, we can't predict or prevent every case.

TC proves emotions aren't just mental; they have real, physical consequences. Treating stress as a whole-body issue is key to protecting your heart.

Stress management/ cardioprotective hack	General heart health benefits
Mediterranean diet	Proven to protect your heart by lowering heart attack/stroke risk (backed by major studies)
Dark chocolate (≥74% cocoa content)	Lowers blood pressure and stress hormones (40g per day shown to be effective in human trials)
Exercise	Lowers blood pressure, strengthens heart muscle and improves blood flow
Psychological help (therapy, mindfulness)	Reduces stress chemicals; lowers heart rate and inflammation (especially during tough times)
Pet ownership	Pets reduce daily stress (best evidence for this relates to dogs)
Oral hygiene (gum health)	Healthy gums = less body-wide inflammation; linked to better blood vessel health and less scar tissue in heart (dental studies confirm)
Regular sleep schedule	Consistent sleep protects blood pressure; night shifts raise heart risks by 10–30 per cent depending on exposure and individual factors

THE FACE
IDENTITY, EXPRESSION AND THE SOCIAL BRAIN

2.

DO WE ALL HAVE A 'RESTING FACE' AND WHAT DOES IT MEAN?

'Come on, try and look happy'

'What's wrong with you?'

'Why the long face?

Have you ever had comments such as those on the right thrown at you and felt completely bemused as to why this complete stranger assumed you were sad, worried or angry? The annoying thing is, you were having quite a good day up until that point. You were sitting there, perfectly at ease, face just resting in neutral mode. Congratulations: you've just been judged by your so-called 'resting face'. So, why is it that some faces at rest send out the wrong social signals?

IS IT REALLY 'RESTING'?

A 'resting', 'natural' or 'neutral' face is the default expression you have when you're not consciously trying to communicate anything or convey a particular emotion. But are your face muscles completely relaxed? From a clinical point of view, a resting face is one that should be electrically silent when muscle activity is measured by an electromyogram (EMG). Any movement or electrical signal detected while your face is supposed to be at rest is a sign that something's amiss with your facial nerves or muscles.

But, from a neurophysiological perspective, your facial muscles should never be electrically silent. When measured with an electroencephalogram (EEG), baseline activity should always be present. Even when you think you're perfectly relaxed, a small subset of motor units (called single-motor units, or SMUs) continue firing at a low, steady rate to maintain the tone of your jaw muscle (temporalis) and forehead muscle (frontalis). This maintains your jaw's resting posture, preventing it from flopping open, and keeps your brow poised even when you're not deliberately raising your eyebrows. It ensures that you can spring into action, chewing, speaking or expressing emotion without delay. So, a 'resting face' is really a bit of a misnomer considering there will always be baseline tension. What does contribute significantly to your neutral expression is the individual structure and muscularity of your face, and this varies from person to person.

MUSCLE TONE

The face that you wear when you're simply being you can be broken down into three areas:

FACIAL STRUCTURE

- Your underlying bone framework sets the basic shape of your face and determines where shadows fall
- Fat pads cushion the bones, creating volume and smooth contours and providing light-reflective volume
- Genetics determine the features you inherit, such as eyebrow position, mouth shape and facial symmetry
- Having a recessed jaw, sunken temples or low brow creates a tired, stern or sad neutral expression

INHERENT MUSCLE TONE

Underlying muscle tone provides a subtle tension that acts like invisible puppet strings to sculpt the features. The degree of baseline activity varies among individuals:

- Orbicularis oris tone (pulls lips inwards) ➡ narrow mouth
- Zygomaticus minor tone (elevates cheeks slightly) ➡ resting smile
- Corrugator tone (depresses brows) ➡ angry/scowling neutrality

MUSCLE VARIATION

There is variation in the types of muscles you have (which fine tune subtle movements), how strong they are, where they attach and how symmetrical they are. This determines how unique your resting face is and explains:

- Why one eyebrow naturally sits higher
- Why you have a resting dimple or dimples, a tiny nose wrinkle or slight corner lift in your mouth

This revelation that your 'resting' face isn't really resting is not that surprising when you think that nobody actually walks around with their mouth agape and an expressionless and slack face that you get when you're fast asleep and truly at rest.

TRY THIS:

Close your eyes and mimic how you'd look during deep sleep (as if you were snoring) and then, maintaining that look, open your eyes – most people don't look like that when they are out and about.

Obviously, there is a level of muscle tone to your face, but beyond this your brain is also actively involved in your facial expression by continuously sending low-level signals, often without your conscious awareness. Neurons in the area of your brain that controls the muscles of your face (facial motor cortex) and in the brainstem facial nucleus (origin of the facial nerve) provide a tonic drive – a constant neural impulse that keeps tiny groups of facial fibres engaged so your face never goes completely slack.

NEURAL CONTROL

Your brain's deep emotion centres, such as the amygdala, insula and anterior cingulate, work hand in hand with rapid automatic pathways that link your hypothalamus to your brainstem to modulate the subtle muscle tone of your face. These behind-the-scenes circuits constantly shape your facial expressions, even before you're consciously aware of how you feel. They tweak that baseline firing of facial motor neurons in real-time, adjusting muscle tension to mirror your internal mood, arousal and interoceptive signals (like nausea or pain) long before you make any deliberate expression.

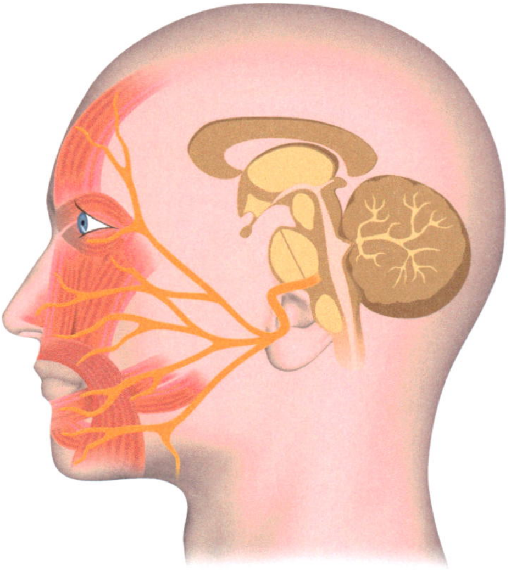

The facial nerve controls your facial movement and expression.

And that's not all – your autonomic nervous system (ANS) also subtly influences how tense your face is based on your internal state. That's because your face feeds back into your body's stress-regulation systems, and this influences whether it appears 'alive' or 'flat'. This happens because branches of the vagus nerve connect to your facial and scalp muscles, and subtle signals from these muscles can activate the nerve's calming parasympathetic pathways. When vagal tone is strong, you settle into a calm, open expression. When it's weak – from chronic tension or stress, say – you default to a restless, guarded or 'mean-looking' face.

As a consequence, continuous stress can turn momentary expressions into permanent resting ones. Each time your sympathetic system spikes, your overall muscle tone increases, including in those tiny facial fibres. Over time, that heightened tension becomes your new baseline, etching a perpetually tense or brooding look into your resting face.

NATURAL SELECTION

It's worth appreciating the role evolution played in shaping our 'resting' faces as social tools: first for survival and now, rather unfortunately, for the unintended judgements of others. Darwin proposed that facial expressions are essentially hardwired biological adaptions that might have been advantageous for our ancestors. The main muscles, such as the zygomaticus major (smiling) and corrugator supercilii (frowning), are present in all of us, and they are highly symmetrical, suggesting there was an evolutionary selection process for clear emotional communication. It's theorized that this could have helped our ancestors maintain cooperation, hierarchy and group cohesion.

Blame your parents for your resting face. Research suggests that it's heritable: if your parents had a naturally stern or friendly neutral expression, you likely will too. Don't believe it? Well, the evidence shows that even those born blind, who have never seen facial expressions, somehow develop family-specific neutral expressions, which proves these traits are biologically ingrained rather than learned.

Three sets of faces showing neutral, angry, and happy expressions (from top to bottom), displayed on a male and female face. The images show how the same expression may be interpreted differently when presented on faces of different genders.

Some resting faces look structurally similar to primitive expressions used to signal emotions. Those with low furrowed brows, for example, inadvertently signal anger, while rounded cheeks denote happiness. Such functional equivalence can lead to snap judgements, even when no emotion is intended. Unfairly, this also continues when resting faces 'signal' disease or poor genetic fitness. Because we evolved to avoid these traits, we tend to subconsciously prejudice faces that suggest illness or anomalies – for instance, faces that are asymmetrical or that droop can unfortunately be perceived as being less trustworthy or competent.

It's a shame people are quick to judge your personality and mood from your resting face, because the evidence shows weak correlation between the two. Culturally, we link femininity with warmth and happiness, and masculinity with dominance and anger. As a result, our brains are tuned to detect happiness more quickly in female faces and anger more quickly in male faces, even when those faces are only subtly smiling or scowling. In laboratory tasks, participants identify happy expressions on women and angry expressions on men faster and more accurately than the reverse.

When a woman's neutral face falls short of the 'always happy' stereotype, observers are prone to label it as anger. That misattribution then triggers harsher social judgements: she looks uncooperative or hostile simply for not looking cheerful – hence the opening comments to this section. At the same time, any face with masculine structural cues (strong jaw, heavy brow) skews even neutral expressions towards anger, so those faces are rated as angrier regardless of the actual emotion displayed.

There is no such thing as a truly neutral face, because every 'resting' expression tells a story about biology, genetics and lived experience. Hopefully, understanding the science will make us all more compassionate interpreters of others' expressions.

Do the following light-hearted activities to decode and resculpt your own resting face (grounded in real physiology).

RESTING-FACE QUIZ AND SCORE INTERPRETATION

For each of the five items below, tick the box that best describes you right now.

	Never	Rarely	Sometimes	Often	Always
Factor	**1**	**2**	**3**	**4**	**5**
Dark circles/under-eye hollows	☐	☐	☐	☐	☐
Jaw or temple tightness at rest	☐	☐	☐	☐	☐
Noticeable 'resting frown'	☐	☐	☐	☐	☐
Forward head posture	☐	☐	☐	☐	☐
Breath holding or shallow breathing	☐	☐	☐	☐	☐

Total score:	

Score Your Face: Give yourself one point if you ticked never, two points if you ticked rarely, and so on. Add your ticks (minimum score = 5; maximum = 25) ➡ Total / 25

5–10 (BASELINE): Your tissues have good tone, collagen support is robust, and low background muscle tension means your 'neutral' appearance reads as relaxed and friendly.

11–17 (SLIGHT PULL): Early signs of volume loss (under-eye hollows) or tension (jaw clench) are nudging your default toward fatigue or mild frown. Small and easy reset actions, like unclenching your jaw can rebalance your baseline.

18–25 (TENSION SET): Chronic stress, posture and volume changes have wired a more serious, fatigued or dominant resting expression. You'll benefit most from targeted release techniques, such as massaging your temple, or slowly stretching the side of your neck to relax tight muscles, and vagal tone boosters, such as slow diaphragmatic breathing, or humming to calm your body.

Volume plus Gravity: If your under-eye area is starting to look sunken or shadowed, it's partly because you've lost some of the natural 'padding' there, and gravity is making the remaining skin sag more noticeably.

Neuromuscular Tone: If you find your jaw or temples constantly feel 'on', that means your facial muscles are never really letting go. What you think of as a 'neutral' face still has background tightness built in.

HABIT INTERRUPTION MATRIX

If you're prone to any of the common expression patterns below, each time you catch yourself in the corresponding trigger situation, immediately perform the interrupt cue. If you do this consistently, you can break the loop of tonic contraction (continuous low-level muscle tightening) that slowly etches an expression into your face.

Common expression	Typical trigger	Interrupt cue	Scientific rationale
Brow furrow	Reading tense email	Exhale and raise brows gently	Breathing out helps calm your nervous system, which relaxes the muscles in your forehead
Lip press	Driving in heavy traffic	Inhale through nose, release jaw	Breathing in naturally helps to loosen and soften your jaw muscles
Jaw clench	Concentration at desk	Place finger between molars, wiggle lightly	Becoming aware of your body's position helps to reset and coordinate your muscle movements
Chin jut (forward)	Phone use, head down	Lift gaze, tuck chin back	Improving your neck posture takes pressure off your jaw joint.

HOW TO TELL IF SOMEONE'S LYING

Picture these two scenarios:

1 Your *forehead sweats* during a job interview question directly linked to an overstated skill of your abilities on your CV.

2 Your *pupils suddenly widen* after your boss asks why you weren't in the office yesterday afternoon after they returned impromptu from a cancelled work trip.

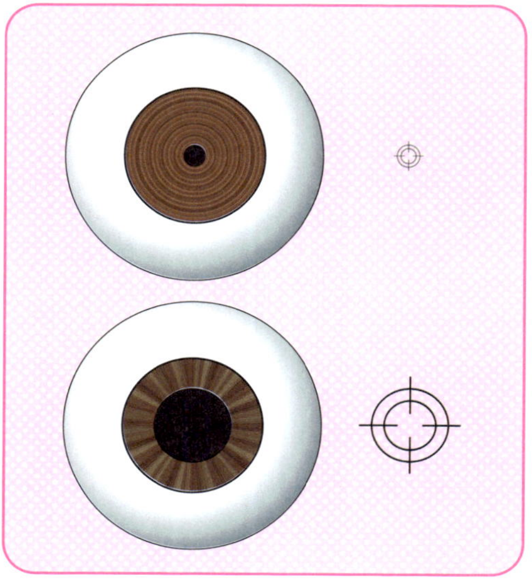

These reactions in your face (that you've no control over) signal more than just simple nervousness; they can, in fact, reveal that you're about to lie.

Although it would be awesome to spot whenever someone is lying, unfortunately (or fortunately) it's not a skill most people have. As humans we detect lies at barely 54 per cent accuracy, which is only marginally better than tossing a coin to decide.

THE SCIENCE BEHIND DECEPTION

To understand how your body reveals your lies, it would help to know what happens in the place where they're first conceived: the brain. Lying is mentally taxing and forces you to juggle many layers of thought. You use up your working memory just to keep track of both the truth and the lie, all the while trying to suppress the truth from slipping out. Then your brain has to task-switch, toggling between different versions of reality – what's real versus what's fabricated.

In 2001, Dr Sean Spence and colleagues from the University of Sheffield published one of the first objective studies investigating brain activity during deception. Using fMRI scans on a total of ten young male volunteers, they found that the regions involved in a wide range of cognitive functions, including working memory, response selection and emotion regulation in the bilateral ventrolateral prefrontal cortices (VLPFC), were significantly activated during lying compared to when telling the truth. Limitations of the study associated with its design and small sample size meant that other areas of the brain could not be excluded, suggesting that deception may rely on a wider neural network than the VLPFC alone. Neuroimaging studies have since shown that when you lie, the frontal and parietal lobes of your brain involved in complex decision-making and self-control light up more than when you tell the truth.

From a moral perspective, the more honest you are by nature, the more mental effort it takes to tell a lie, and this internal struggle is visible in brain scans. Interestingly, for habitual liars, the situation reverses: their brains show increased effort when they must tell the truth. This is because your brain expends more energy when it has to use a pathway that is less familiar. Over time, it essentially builds 'routes' that reflect your moral habits, making lying or honesty either easier or more mentally taxing depending on your patterns.

With this in mind, those regions activated during deception could actually be a reflection of the heightened cognitive load and effort required to maintain a lie. In other words, neuroimaging techniques might just be detecting the effort of cognitive control and conflict required rather than any deception. This understanding underscores the real likelihood that there is no singular 'lie centre' in the brain; rather, deception is distributed in a network of cognitive processes that are engaged in a more demanding way when you deviate from your default behavioural pattern. This has real implications for any neuroimaging-based lie detection.

Lying is a bit like going to the gym because it makes you sweat – not surprising given that it gives your brain a good workout. Your body's autonomic nervous system (ANS) responds to the lie by ramping up sweat production – something that can be measured through skin conductance – as well as altering your heart rate and breathing patterns. But these changes aren't exclusive to lying; they commonly appear whenever you're feeling anxious, conflicted or under pressure. They serve as general 'danger' alerts rather than indicators of deception.

Such brain strain and ANS arousal manifest as observable bodily 'leaks', but they're easily misread. Even if someone fidgets, sweats or avoids eye contact, it doesn't necessarily mean they're lying, as it could just be stress or discomfort – context is key.

A LIAR'S TELL-TALE SIGNS

First in line for potential tell-tale signs are the ever popular facial cues. Micro-expressions, such as that fleeting smile, quick raised eyebrow or slight twitch of the mouth, despite the hype, are not backed by consistent peer-reviewed science as being indicators of a lie.

Physical behaviours are also a bit of a mixed bag. Pitch changes in voice lack robust data, but meta-analyses suggest that liars often pause for longer when asked a question, possibly because of the increased cognitive load. The Spence study mentioned previously also observed this 'vocal cue' in their test subjects, although it was conducted in a highly constrained experimental setting, where subjects merely responded with opposites rather than generating novel lies. The correlation is there, but it's weak.

Physiological markers, which reflect involuntary bodily reactions, offer more reliable clues to deceptions. Skin conductance level (SCL) is the gold standard, indicating stress-driven autonomic arousal that is difficult to control consciously. Pupil dilation is an emerging deception marker that, when combined with SCL, can improve detection accuracy. Research, including studies by Kircher and colleagues in the mid-2000s using the Ocular-Motor Deception test (also known as Eye Detect), highlights the promise of SCL and pupil dilation as practical tools for identifying lies. Other measures, such as monitoring heart and breathing rate or facial temperature, show low or inconsistent reliability.

The Ocular-Motor Deception Test (ODT) uses eye tracking to detect signs of lying by measuring small changes in response time (RT), eye movements and pupil size. In this study, which compared neutral statements with crime-related 'cash' questions and a plausible but uncommitted 'exam' theft alternative, guilty participants generally took longer to respond than innocents, except on money-related 'cash' questions, where they reacted more quickly because the word was tied to what they were hiding. This shows how guilt can subtly alter response speed, though the test still faces limits and is best used alongside other methods.

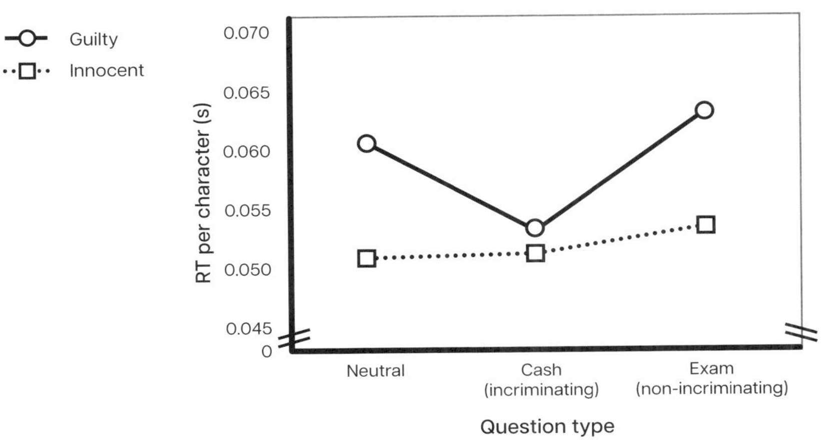

Contrary to popular opinion, body movements such as fidgeting aren't considered to be validated indicators of lying. Surprisingly, research suggests liars may in fact move less, not more, which counters the 'twitchy villain' trope. This pattern was found robustly in a 2010 University College London study that sought to explore whether spontaneous bodily motion could reveal when someone is lying, without relying on traditional cues such as facial expressions or verbal inconsistencies. It suggested that motor suppression (a kind of behavioural freeze in which liars lock-down their behaviour, gestures and movements, consciously or unconsciously, to avoid giving themselves away and appear unnaturally still) might be a more reliable signal of deception. However, the scientists cautioned against calling it a conclusive 'sign of lying' because lying is complicated. People's thoughts and emotions change depending on the situation, and not everyone acts the same way when they lie.

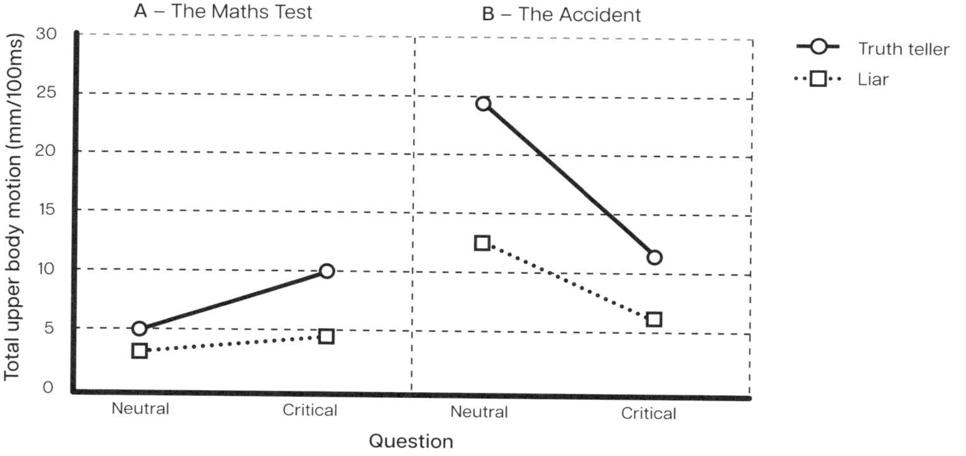

Participants answered neutral (unrelated) and critical (related) questions after staged maths/accident deception scenarios. Motion tracking revealed reduced body movement in liars, especially in response to critical questions. Truth tellers generally moved more than liars in both situations. Liars tended to stay more still, especially during the accident questions.

THE PROBLEM WITH POLYGRAPHS

For a topic focused on detecting lies, it might seem odd to have gone so far without even mentioning polygraphs, but that's because they are scientifically unreliable. You typically see polygraphs, or lie-detector tests, which measure physiological responses such as heart rate, blood pressure and breathing while a person answers a salvo of questions, in crime movies. However, there are a number of limitations with using them:

1. Physiological responses vary widely from person to person – without knowing a person's resting heart rate, for example, any changes are meaningless.

2. They measure arousal, not lies, via the tracking of physiological changes such as sweat, heart rate and breathing rate, but these spike for any strong emotion: anxiety, fear, excitement, even concentration. They can, therefore, identify innocent people as deceptive (false positives) simply because they're stressed.

3. Machiavellian/psychotic individuals tend to feel less guilty about things, which means they'll be less physiologically aroused when they lie.

4. There are no real universal deception cues, because culture can influence a person's physical responses. Some people have learned to smile when they're anxious, while in some cultures averting one's gaze can signal respect and not deceit.

5. People have learned how to 'game' the test: biting your tongue, doing mental arithmetic or even tensing muscles can dull the contrast between lie and truth signals. These tricks scramble the baseline response, making lies harder to spot.

6. They're legally dubious. In the US, under Frye and Daubert standards (rules that guide what counts as reliable scientific evidence), polygraphs fail the 'general acceptance' test. Most courts in the world won't admit them as legitimate proof because they're considered too unreliable and controversial.

Armed with this knowledge, how would you go about catching a liar? Some polygraph experts maintain that accuracy can go up to between 80 and 90 per cent when there is a well-trained examiner who knows how to conduct the interview and the test is carried out with proper quality controls in place, though accuracy averages at 56 per cent.

The technique known as cognitive interviewing (asking unexpected, mentally taxing questions, such as 'Can you describe the event backwards?'), reportedly increases the accuracy to 76 per cent because it overloads a liar's working memory.

In closing, if you want to know if someone is lying to you, prioritize open dialogue. By replacing suspicion with curiosity – 'Help me understand why it's not adding up' versus 'You're lying!' – you're more likely to catch them out because suspicion shuts people down, while curiosity keeps them talking, and deception is far easier to detect when someone is still engaged in dialogue. Also, looking for a combination of indicators rather than a single one helps reduce the likelihood of mistakes and offers a more accurate, science-based approach to spotting a liar.

ARE YOU WEARING SOMEONE ELSE'S FACE?

We all have that one celebrity lookalike, the person who could star as us in a film about our lives. Yet when you come across a complete stranger who is nearly your exact mirror image, it sparks a different kind of wonder. It makes you question: could they actually be a distant relative, or is nature simply playing tricks with our faces?

Doppelgangers are defined as individuals who possess nearly identical facial features despite having no familial relationship. Sometimes called ghostly doubles or lookalikes, the term doppelganger is a literal translation of 'double walker' in German.

According to folklore, encountering your doppelganger was considered an omen of bad luck or even a sign of impending death. But this hasn't stopped Canadian photographer Francois Brunelle, who since 1999 has brought together 250-plus pairs of doppelgangers to curate a worldwide collection of their portraits for his 'I'm Not a Look-Alike!' project.

How is it that genetically unrelated individuals share near-identical faces? Is random chance at play here, or is some deeper, biological process underpinning this phenomenon?

The probability of encountering a real-life, exact doppelganger is extremely low, with some studies suggesting it's as unlikely as one in a trillion.

A 2022 study in *Cell Reports* (in collaboration with Brunelle) revealed an intriguing pattern: genetically unrelated lookalikes share similar DNA sequences for facial features, body type and even behaviours such as smoking. But, beneath these similarities, key biological differences emerge, particularly in how their genes are expressed and the makeup of their gut microbiomes. This finding challenges the classic nature versus nurture debate, showing that human individuality stems from the interplay of genetics and environmental influences and not genes alone.

THE ROLE OF GENETIC VARIATION

The main reason unrelated people can look extremely similar is because of their DNA, specifically in certain types of genetic variations. Cue the question: why is the term 'variation' used when referring to lookalikes who actually share similar DNA? To understand this, you need to appreciate how geneticists talk about similarity and difference.

Genetic variation essentially describes diversity in the human population. About 99.9 per cent of our DNA is identical identical to everyone else's DNA. The remaining 0.1 per cent (approximately 3 million positions or locations on our DNA) is where genetic variation occurs, which is what makes us different. So, the term 'variation' refers to differences across the entire human population, not necessarily differences between a specific lookalike pair.

Single nucleotide polymorphisms (SNPs, pronounced 'snips') are the most common type of genetic variation between people. They represent a difference in a single building block (nucleotide) at a specific location in the DNA sequence – think tiny spelling differences in the genetic code.

So, for example, take 'region 21' on a gene, which is a stretch of DNA that contains instructions on how to build specific proteins or molecules:

Person X might have a 'C'

Person Y might have a 'T'

Person Z might have a 'C'

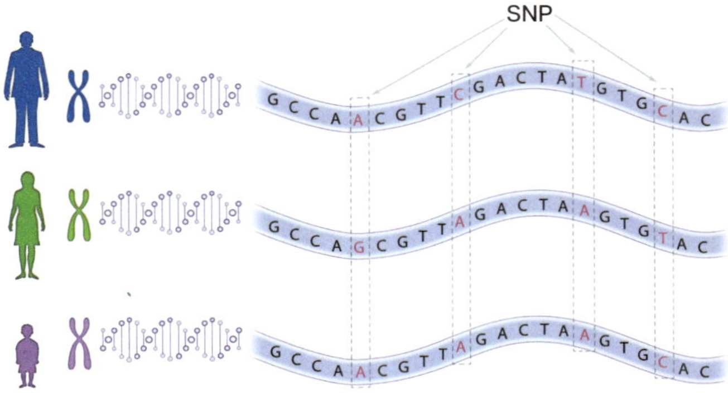

The fact it's possible that a different letter (nucleotide) can be in region 21 is the genetic variation – it basically says this is an area where people tend to vary (can be different). So, the term 'variation' describes the potential for difference across all humans at these spots. Lookalikes, on the other hand, happen to be the same at a very large number of these specific variable spots (especially regions known to affect facial features).

Using advanced facial recognition algorithms and 3D imaging software, the researchers of the Cell Reports study previously mentioned were able to meticulously screen large numbers of people and identify those who matched a strict standard of similarity. Unrelated people who got scores as high as real identical twins were flagged as 'ultra' lookalikes, and 3D face maps that

THE BUILDING BLOCKS OF DNA

A DNA nucleotide consists of three parts: a deoxyribose sugar, a phosphate molecule that forms the backbone of the chain, and one of four nitrogenous bases – adenine (A), thymine (T), cytosine (C) or guanine (G) – that determine the type of nucleotide. Because DNA is a double helix (like a twisted ladder), the bases of each nucleotide form its rungs and follow strict pairing rules to keep the DNA stable: C always bonds with G and T always bonds with A.

scanned and measured their facial features provided objective, quantitative data on the similarity of their physical characteristics.

As a result, the researchers were able to identify and analyse 16 pairs of 'ultra' lookalikes, finding that more than half of them shared an unusually high number of the same common SNPs (identical DNA letters): nine pairs shared 19,277 identical SNPs located within 3,730 different genes. In other words, these doppelgangers don't differ from each other at these specific sites even though the wider population typically does. Some people will match at a few of these sites, but it's rare for people to match at so many locations important for facial features.

Twin studies have shown that more than 70 per cent of the reason why people have different face sizes, nose shapes and eye spacing is due to their DNA. In the doppelganger study, half of the unrelated pairs (16 out of 32) were so similar that three different face-comparing computer programs gave them the same similarity score as actual identical twins. Because all three programs gave the same result, this proved their resemblance was an objective reality, not just a subjective opinion.

The similarities between them went beyond just their faces. These doppelgangers also appeared to have similar physical and behavioural traits.

TRAITS

Meeting someone who looks exactly like you is freaky enough, but the thought that they might also walk like you or even have the same peculiar half smile would be enough to make you think you'd entered the Twilight Zone.

Well, it turns out that doppelgangers often share more than just their faces. The same research found that these pairs tend to be similar in height, weight and overall build. Even more remarkable was their often shared smoking histories (either both smoked or neither smoked) and similar education levels. How could this be so?

At the heart of this mirroring is pleiotropy, a genetic principle where one gene influences multiple seemingly unrelated traits. Those shared genetic variants that make faces look alike might also influence other physical traits such as height and weight.

Take the PAX3 gene, which is crucial for the development of the neural crest, a group of cells in the early embryo that gives rise to a huge range of structures, including facial bones and cartilage, pigment cells and parts of the nervous system. Mutations in this gene are known to cause Waardenburg syndrome, which leads to wide-set eyes, a white forelock and different-coloured eyes. This shows that one gene can have diverse effects simply because it's involved in a fundamental development process that influences many downstream traits. It is plausible, then,

that the same genes that make doppelgangers look alike might affect not only their facial structure but also other aspects of their bodies, and could potentially influence their behaviours.

Unlike identical twins, who are genetic clones raised in the same environment, doppelgangers might start with similar genetic blueprints for a face or frame, but their life experiences, environments and remaining genes send them down vastly different biological and behavioural paths.

LIFE GETS IN THE WAY

Despite looking extremely alike and sharing many genetic variants, just one pair out of the sixteen ultra lookalike pairs studied had similar communities of microbes in their mouths. This finding shows that while doppelgangers can be genetically similar and share physical and behavioural traits, their microbiomes (the collection of bacteria, viruses, fungi, etc. living in/on them) can remain distinct. This difference is significant because if lookalikes did share microbiomes, it would suggest shared health risks or resilience and imply a stronger genetic pull on microbial composition than currently accepted. It would even complicate the use of microbiome forensic 'fingerprints', which are considered highly individual. This dissimilarity was evident even when lookalike pairs had similar body weights, proving that weight similarity alone doesn't guarantee microbiome similarity.

Your diet, where you live, what drugs or supplements you take and your daily habits each carve out a unique microbial community. In fact, one major study by Dr Rothschild and her colleagues in 2018 found that more than 20 per cent of microbiome differences come from environmental factors. Your genes? They explain less than 2 per cent. Genetics plays a minuscule role compared to environment.[2]

---

[2] One caveat is that the 2022 doppelganger study looked at the oral microbiome only, which might not reflect what was happening in other parts of the body, such as the skin and gut. So, while the findings were compelling, they did not necessarily provide the whole picture.

Think of your microbiome as a living record of your life. These microbial differences act like 'biological signatures', revealing your distinct history of food, travel, stress and more. They help explain why doppelgangers, despite shared genes, are never perfect biological clones: the environment sculpts each person's microbial identity.

Although the 2022 doppelganger study was a significant one, it was limited by its small sample. With more people, it would have had a better chance of spotting true effects and connections. The fact that the doppelgangers were predominantly of European ancestry (13 out of 16 pairs) limited the generalizability of its conclusions further.

What's clear is that, although genetics might create the template, it's the environment that writes the details. You could be 'sharing' someone else's face, but that invisible microbial and molecular signature of your life is all yours.

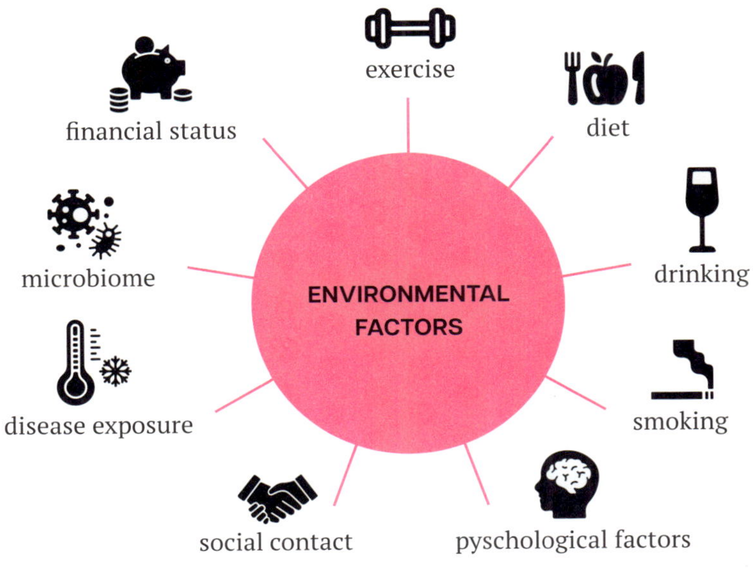

WHO'S THE DADDY?

It's often said that newborns tend to resemble their fathers, at least in the earliest weeks. While babies might gradually take on features from both parents as they grow, this early dad-likeness is well-known parenting folklore. One popular theory suggests that it serves an evolutionary purpose, helping to reassure dads of their biological connection to the child and encouraging their involvement from the very start. The evidence for this is inconsistent, especially given that some children are actually the spitting images of their mothers.

Baby-swap scenarios aside, fathers don't, unlike mothers, have absolute certainty that a baby is theirs. And with the rise of easy-to-access, direct-to-consumer (DTC) DNA and ancestry testing kits, there's been a surprising consequence: estimates suggest that as many as 5 per cent of users worldwide have discovered unexpected paternity results, potentially impacting millions of families.

DTC genealogy has exploded over the past decades, with more than 30 million people around the world estimated to have taken tests, and a global market now worth around $2–3 billion annually for at-home kits from services such as Ancestry, 23andme and MyHeritage (they can cost anything from £79 in the UK, and from $99 in the USA for basic ancestry tests, with higher prices for premium or health-focused bundles). But, beyond ancestry charts, these tests are unearthing biological truths that rewrite family narratives overnight.

Nowadays you don't have to wait for the baby to be born to find out who the father is. Non-invasive prenatal paternity (NIPP) tests analyse foetal DNA found in the mother's blood, and with 99.9 per cent accuracy and zero risk of miscarriage, they offer a safe and accurate method during pregnancy to determine the biological father of a foetus.

PATERNITY TESTS

What exactly happens when you take a test to determine who the father is? Paternity tests look at tiny stretches of DNA called short tandem repeats (STRs), which represent specific locations on your DNA where a very short sequence of bases (say, AGAT) repeats back to back. The number of times this sequence repeats varies between different people. Because you inherit half your DNA from mum and half from dad, this means, for each STR location, one of your repeat counts must match one of mum's counts and one must match one of dad's counts.

Polymerase chain reaction (PCR) technology (a DNA-copying method) has made DNA testing possible from tiny samples, enabling highly accurate and reliable paternity results. Modern labs use STR profiling to distinguish your genetic profile from others with far greater precision than older methods, increasing certainty and reducing errors. Legal paternity tests require strict procedures for admissibility, while cheaper home kits are suitable only if you want answers for your own peace of mind.

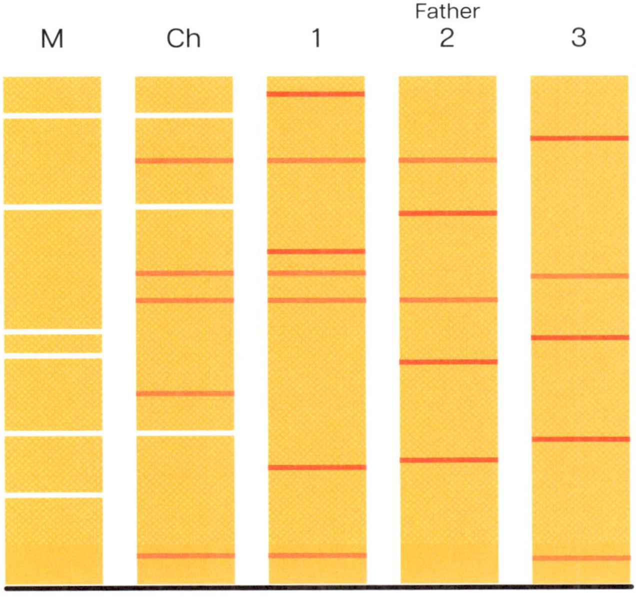

DNA paternity test: Genetic fingerprinting results. Samples obtained from the mother (M), child (Ch) and possible fathers. The DNA bands in the child's sample should be a combination of the mother's and father's samples. Alleged father 1 is the child's biological dad in this example.

STEP-BY-STEP PROCESS (SIMPLIFIED):

1. Collect cheek swab samples from child, mother and alleged father.
2. In the lab, isolate DNA and target 15 to 20 STR sites spread across the genome.
3. Use PCR to amplify each STR region, then measure the repeat counts.
4. Compare the child's STR profile to both parents' profiles:
   - At each STR site, one of the child's two repeat counts should match mum's and the other should match dad's. Testing 15 to 20 independent DNA markers significantly reduces the chance of an accidental match, because a consistent match across all tested sites indicates a greater than 99.99 per cent certainty who the biological father is.
   - If the alleged father's profile fails to provide the non-maternal half at two or more STR sites, he is 'excluded' as the biological father (100 per cent certainty).

WHEN YOU'RE NOT THE FATHER

Imagine finding out that the child you had loved, cared for and nurtured over the years was not biologically yours. This has been the reality for countless men throughout history and across diverse cultures, but the truth of biological parentage has often remained obscured until the advent of modern DNA and ancestry testing.

Extra-pair paternity (EPP) is the technical name given to describe when the biological father of a child is someone other than the mother's current social partner (husband, boyfriend). This designation discounts adoption and known out-of-wedlock births and refers only to cases where a social father is unaware of misattributed paternity.

The EPP rate is the percentage of children born within the aforementioned partnerships that are genetically fathered by an extra-pair male. Large-scale reviews (meta-analyses) consistently show EPP rates are surprisingly lower than expected, typically ranging between 1 per cent and 3 per cent in contemporary Western societies, contrary to long-standing widely cited claims that it was around 10 per cent.

Rates vary dramatically across human societies and historical periods. A groundbreaking 2020 study by Professor Brooke Scelza and colleagues reported an extraordinarily high EPP rate of 48 per cent among Himba cattle farmers in Namibia. However, this high rate is largely down to cultural norms and practices that permit women to have second husbands.

Belgian geneticist Dr Maarten Larmuseau combined historical records with genetic data (Y-chromosome DNA tests) to reveal that EPP rates were notably higher in poor urban areas of 19th-century Western Europe (around 6 per cent) compared to more rural communities during the same era, where rates hovered between 1 per cent and 2 per cent. This disparity underscores how social environments – in this instance, anonymous cities versus close-knit rural areas – can significantly influence EPP rates.

From an evolutionary perspective, you might wonder why EPP occurs at all. Some biologists suggest that women may seek out 'good genes' (traits such as disease resistance or physical attractiveness) for their children by choosing men considered more genetically superior to their social partner. For men the evolutionary motivation can be quite different: it could allow them to father more children without making the substantial commitment that active parenting requires. In both cases, these strategies reflect the complex, sometimes hidden forces that shape human behaviour and family lineage.

But here's a thought: what if there's been no cuckolding of any kind, and there's no question that the child is yours, but the highly accurate and non-faulty DNA test says with 100 per cent certainty that your and your child's DNA absolutely do not match.

ARE YOU MY FATHER-UNCLE?

In such circumstances, where immaculate conception is not an option, there can only be one other explanation for all this: you're a chimera.

A chimera is someone who has cells that carry two distinct sets of DNA (two genomes). It most commonly occurs when non-identical (fraternal) twin embryos fuse very early on in pregnancy or when one dies and their cells are absorbed by the surviving twin (vanishing twin), resulting in a single baby that's made up of cells from both original embryos. Some tissues derive from twin A, and other tissues derive from twin B. As a result, DNA taken from blood or cheek swabs (used in routine tests) might not match the DNA present in germ cells (sperm or eggs) that are passed on to your children. And therein lies the problem.

In 2015, a couple from Washington state in the USA who had conceived a child via IVF were stunned to learn that there was a blood-type mismatch between them and their baby. This prompted a DNA test that revealed the father was not the biological parent, with the test showing only 25 per cent genetic relatedness (typical of an uncle–nephew relationship), not the expected 50 per cent. It was only when they tested his semen, which surprisingly revealed a different DNA profile, that they obtained a 50 per cent match, proving he was indeed the biological father. Geneticist Barry Starr identified the man as a chimera who had absorbed cells from his fraternal twin in utero via 'vanishing twin syndrome' (estimated to occur in roughly 15–35 per cent of twin or multifoetal pregnancies, though rates vary depending on how and when measured). Cheek-swab DNA (used in the test) reflected his genome, but his sperm cells contained his twin's genome – meaning he had fathered the child using his deceased brother's DNA.

This remarkable biological paradox meant that the child was, genetically, the man's nephew – the son of the twin he had absorbed in the womb. While he was the child's social and legal father, he was, from a DNA perspective, the biological uncle. Most of his sperm carried his own genetic material, but about 10 per cent carried his twin's DNA, which explained why his other biological child matched his own genome.

Cases of tetragametic chimerism (two embryos fusing), severe enough to cause obvious discrepancies in standard DNA tests like the example above, are extremely rare, with only about 100 confirmed cases reported in the medical literature. But if chimerism is suspected following a bizarre DNA test result, the solution is multi-tissue sampling. Testing DNA from different tissues (for example, blood, semen, hair follicles, skin biopsies, cheek (buccal) swabs from different areas of the mouth) can reveal the hidden genome. SNP microarrays (genotyping chips – small glass or silicon slides covered with DNA probes that let scientists scan hundreds of thousands of genetic variants at once) are particularly useful, as they can detect the presence of mixed cell populations and identify multiple distinct DNA profiles within one person.

DTC genetic tests have the potential to unleash a wave of paternity revelations that leave individuals grappling with their identity and family lineage. This has led to the emergence of non-paternity-event support groups. And to resolve ambiguous or conflicting genetic results, laboratories are increasingly turning to next-generation DNA sequencing (NGS) rather than relying solely on traditional STR tests. This newer approach not only improves accuracy but also raises important ethical questions that extend beyond revealing a person's identity to uncovering hidden health risks. Consequently, genetic counsellors (specialized healthcare professionals who act as expert guides and a support system for individuals and families concerned about genetic conditions) are tasked with the unenviable role of balancing an individual's right to know against the potential for family disruption.

AM I BEAUTIFUL?

Take a close look at your reflection in the mirror. Does the face staring back at you generate a 'yes', 'no' or 'maybe' answer to the title's question? Now ask yourself – why did you come to that conclusion? Is it because that's what you've always been told, or are you being overly kind or self-critical? What's it about your face that makes you think you're beautiful, unattractive or anything in between?

The truth is, many hidden evolutionary forces have helped shape how you judge your own beauty. While beauty standards vary dramatically across cultures, from the lip plates of Ethiopia's Mursi people to the 'glass skin' ideals of Korea, they share universal evolutionary roots that are deeply ingrained in survival and reproduction. The unifying thing about beauty as a concept is how it echoes our deep, ancient instincts: our ancestors chose partners whose features hinted at health and vitality (for that particular time and environment). However, these primal cues still inform today's beauty ideals, often in ways we don't consciously realize.

JUDGING BEAUTY

In 2025, Miss Opal Suchata Chuangsri from Thailand was crowned Miss World, and Mr Juul Messiaen of Spain was crowned the winner of the Man of the World competition. But are they beautiful?

Your fusiform gyrus – an outward folding in the brain's cortex involved in recognition – has already answered; within just 0.1 seconds: faster than a blink, your brain makes a snap judgement. Initial reactions to beauty often kick in before we even notice them, but truly deciding whether something is beautiful takes extra processing in the brain's cortex.

Whether or not you agree with the politics of beauty pageants, these competitions have been going on since the late 19th century, and science may help to explain why their legacy continues. Studies show that judging beauty activates the brain's reward system, particularly the orbitofrontal cortex (OFC), a region that also responds to other rewarding experiences such as tasty food. By triggering dopamine, the feel-good neurotransmitter, this reinforces the experience and makes you want to seek out similar stimuli, again and again.

In the Stone Age that began roughly 3.3 million years ago, it was simple: mate choice equalled genetic survival. Our visual cortex was hardwired to spot healthy partners, and a lot of judging went on because facial attractiveness was used to predict immune strength.

	Cue	Link
Health indicators	Facial symmetry	Like a genetic spell-checker, facial symmetry signals developmental stability, indicating that an individual is able to grow without any major genetic or environmental disruption. Such steadiness often points to a resilient body and a strong immune system.
	Clear skin	In pre-historic times, clear skin is thought to have correlated with parasite resistance. High oestrogen can subtly enhance skin's appearance and overall facial attractiveness
Fertility signals • Women	Low waist-to-hip ratio (WHR)[3] ≈ 0.7	Women with low WHR have higher pregnancy rates, making it an 'honest signal' of reproductive health. (High WHR > 0.8–0.85 is linked to lower pregnancy rates.)
	Body symmetry	Associated with regular ovulation.
	Strong jaw	Correlates with testosterone and indicates sexual maturity.
• Men	Body symmetry	Predicts sperm quality.
Youthful features in adults (neoteny) • Women	Large eyes, fuller lips, softer contours	These feminine facial features are often preferred because they signal reproductive health. They're shaped largely by male selection over generations and correlate with higher oestrogen levels and fertility.
	Healthy hair	Reflects nutritional status and hormonal balance.
• Men	Average face	Features that appear overly childlike can conflict with signs of maturity, so female-driven evolutionary pressures over time have favoured male faces that strike a balance: youthful but clearly adult. This mix tends to signal trustworthiness and nurturing potential, unlike very masculine traits, which can sometimes raise doubts about approachability.

[3] If someone's waist measures 70cm and hips are 100cm, their WHR = 70 ÷ 100 = 0.7. This means the waist is about 70% the size of the hips – a body shape often linked in research to health advantages and, in many cultures, perceived attractiveness.

THE UNIVERSAL LANGUAGE OF BEAUTY

If you ask someone what makes a person naturally beautiful, most will say it depends on individual taste, which is often shaped by culture, ethnicity, age, religion, upbringing and personal experience. Yet, despite these differences, certain beauty standards seem to show up across cultures time and again.

These universal preferences stuck around because they helped our ancestors choose healthy partners to pass on their genes. Over time, they became fundamental drives hardwired into our biology that still shape what we find attractive today.

THE IMPORTANCE OF BEING SYMMETRICAL

If you were to rank universal beauty standards, 'symmetry' would likely be near the top. Studies suggest humans are naturally drawn to symmetrical faces and bodies, and those with more symmetry often report greater mating success, though the strength of this link varies across studies and populations.

In environments with high exposure to disease, people tend to develop stronger preferences for traits like facial symmetry and other visible signs of health, because these cues may help to signal a robust immune system and reduce the risk of infection from close social contact. This explains why a study found that, among the Hadza hunter–gatherer people of Tanzania, pregnancy and breastfeeding seem to sharpen a woman's natural 'health radar'. During these vulnerable phases, mothers in such environments develop a heightened sense of who's healthiest and most supportive, both genetically and socially, to ensure the best care for them and their babies (as demonstrated by their preference to be around people with high facial symmetry).

Infant studies offer compelling evidence for an inborn 'beauty filter'. Researchers have noticed that even very young infants tend to look longer at faces that adults call attractive: those that are symmetrical or close to an 'average' prototype. This early bias hints at an innate sensitivity to certain facial patterns, though it's still up for debate whether babies simply find those faces easier to process or if there's an evolved instinct to spot health cues.

It's important to remember that these insights come from lab studies using static photos. Real-world attraction is richer and more dynamic, weaving together facial movement, tone of voice and the context in which people connect. And while extreme asymmetries are often seen as signs of developmental issues and disease, research suggests that prenatal exposure to modern substances like alcohol or retinoids, and even habitual sleeping position, can contribute to facial asymmetries, alongside genetics. This weakens symmetry's reliability as a universal health signal.

THE GOLDEN RATIO

This mathematical model, widely recognized in aesthetics and denoted by the Greek letter phi φ (~1.618), describes a proportion considered especially pleasing to the eye: it occurs when a line is divided so that the ratio of the whole line to the longer segment is identical to the ratio of the longer segment to the shorter one. It's believed to exist throughout nature and is supported by some studies as a universal standard of beauty. Proportions of facial features, such as the ratio of mouth width to nose width, the spacing of the eyes and overall face length, are thought to frequently align with the golden ratio in faces deemed most beautiful.

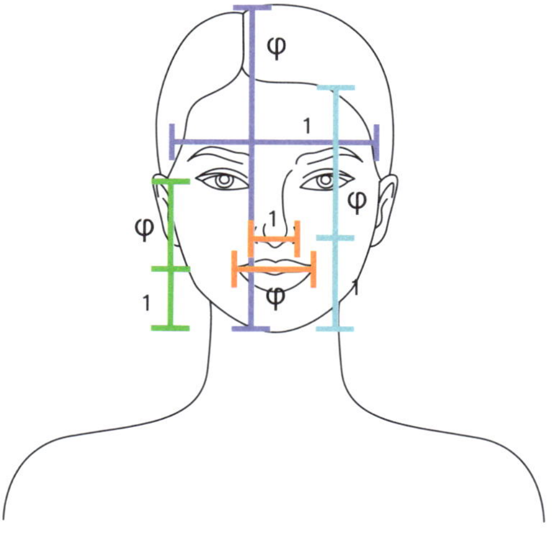

The fact it recurs in geometry, Renaissance art and architecture, and within structures and patterns considered beautiful has led many to believe that it deserves the title 'the universal standard of beauty' and one of the most celebrated constants in both science and art. However, this remains a topic of debate. Some argue that there is no strong evidence to support the claim that the golden ratio inherently makes things more beautiful or aesthetically pleasing. They contend that ideal proportions alone cannot fully capture the complexity and nuance of what it means to have a beautiful face. Furthermore, anthropometric studies (which measure human body dimensions) expose a Eurocentric bias, as many African, Asian and Indigenous facial features do not conform to these standards.

WHY SOME CULTURES PRIZE DIFFERENT FEATURES

Most ideas of facial beauty come from a culture's own story – its history, environment, values and power structure, rather than from some universal, hardwired ideal. Yes, traits such as symmetry or clear skin can signal health and may have an evolutionary basis, but the vast majority of what any society prizes in a face is borrowed from its cultural narrative. Take Maori culture, for example – ta moko tattoos refer to ancestry, rank and tribal identity rather than denoting fertility.

Ultimately, faces become symbolic canvases on which biological adaptations, colonial history, spiritual beliefs and social hierarchies all mingle, making beauty richer and more varied than a simple set of universal evolution-driven rules.

Beauty industries often exaggerate our natural preferences for certain traits such as symmetry and youth, encouraging extreme ideals that can harm self-esteem and distort perceptions of attractiveness. And marketing and digital filters promote narrow standards that overlook real diversity, often leaving individuals dissatisfied and disconnected from authentic and varied human beauty.

So, to answer the question 'Am I beautiful?' remember that beauty was forged primarily by evolutionary pressures to spot health and vitality, not flawless, filtered perfection. But true beauty comes not from traits honed by natural selection – traits shaped by shifting environmental and cultural pressures – but by the unique stories, strengths and quirks that make you unmistakably you. Embrace your own diverse reflection – it's the only standard that really matters.

THE GUT

HEALTH, IMMUNITY AND THE LOGIC OF INSTINCT

3.

COULD YOUR GUT BUGS BE CHOOSING YOUR FRIENDS?

The idea of little microbes in your belly controlling your brain sounds like the stuff of science fiction, but it's quite common in nature. When the fungus *Ophiocordyceps unilateralis* – better known as the zombie-ant fungus – infects an ant, it grows inside its body and eventually hijacks its nervous system. It then releases chemicals that control the ant's behaviour and somehow compel it to climb to a precise location, often a high point on a plant, and bite down in a death grip that locks it in place. After several weeks, the fungus erupts from the ant's head and releases its spores to infect as many other ants as possible.

Now, I'm not saying that your gut microbes want you to clamp your jaws onto the nearest sycamore tree for their own personal gain. Unlike the zombie-ant fungus, which is the archetypal parasite (and undeniably quite a sinister one), your gut microbes tend to work in a more symbiotic way and generally want what's best for you.

YOUR GUT MICROBIOME

There are trillions of micro-organisms in your gut – about 100 trillion to be exact, but that's only an estimate. Most of the inhabitants are bacterial in origin (90 to 95 per cent), with the remaining portion made up of fungi, archaea (another type of ancient microbe), viruses, bacteriophages (viruses that only infect bacteria) and single-celled organisms known as protists. They all play various essential roles to keep you healthy and alive; for example, helping you to digest and metabolize food, fine-tuning your immune system, and preventing you from being infected with opportunistic pathogens.

HOW DID THEY GET THERE?

Most of your first gut bugs are transmitted from your mother, through natural birth (from exposure to microbes in the vagina or rectum), breastfeeding (bacteria and prebiotics in milk) and postnatal skin-to-skin contact. On average, about 50 per cent of your gut bugs come from mum, but this varies depending on factors such as whether you were a natural delivery or C-section baby and breast or bottle fed. Although microbes passed from the mother often dominate a baby's first microbiome, they aren't permanent, because your microbiome evolves with diet, and antibiotic and environmental exposures.

You can think of your gut as a bustling microbial metropolis, where every social interaction becomes a new migration route for its tiny citizens. And throughout your life you constantly exchange your gut strains with the people around you.

Whether a microbe is exchanged between people depends on its ability to survive outside of the body. Strict anaerobes such as *Roseburia* hate oxygen; even a few minutes in air kills them. Aerotolerant microbes and spore-formers such as *Clostridium difficile* are tougher. They can hang out for days to months on surfaces, dust or the objects we touch

How microbes move between us

Primary routes	Examples
Direct physical contact	Handshakes, hugs, kissing, sexual contact and caregiving (e.g., infant feeding) directly transfer saliva, nasal, skin or faecal microbes.
Shared food and objects	Eating and drinking from the same utensils, dishes and bowls, sharing cigarettes, and co-sleeping leads to the exchange of saliva or faecal microbes. Friends sharing meals develop more similar gut communities.
Environmental reservoirs	Surfaces (doorknobs, toilets), water, soil and animals (pets or livestock) can act as intermediaries.

(called fomites). *C. difficile* is a harmful bacterium you want to avoid as much as possible; if you get infected it can cause severe inflammation in your colon, which in some cases can lead to serious complications such as sepsis. Following basic hygiene practices including washing your hands with soap and water rather than using just hand sanitizer (which doesn't seem to affect it) can keep it at bay, but, unfortunately, it's not easy to kill and therefore is highly contagious. It survives by forming tough, protective shells that help it resist harsh environments such as heat, dryness and disinfectants until it can reactivate.

After entering the mouth, a microbe has to contend with even more obstacles before it can colonize a new gut. If it passes through the stomach acid and digestive enzymes intact it then needs to find itself a niche within its new home – that is, a spot where it's not immediately outcompeted or attacked. This depends on the diversity of other strains in the recipient's gut environment and the strength of their immune system.

SOCIAL INTERACTION

Studies reveal that people in close social contact share significantly more gut microbial strains than unrelated individuals or those living apart. In 2025, the journal *Nature* published findings from a study tracking social connections and gut bacteria in nearly 1,800 adults across isolated Honduran villages and found the strongest sharing of gut bacterial strains occurs between spouses, cohabiting partners and mothers with their children. Married couples, for example, shared about 12 per cent of their gut bacterial strains – roughly double the rate seen between unrelated villagers who didn't live together. This similarity persists even when accounting for shared diets.

The extent of closeness between people matters too: in a US study of 94 spouse pairs sampled in 2014 and 2015, partners reporting being 'very close' showed exceptional microbiome overlap of 30 per cent on average. In the Honduras study, even greetings like cheek-kissing were linked to higher strain sharing (around 13 per cent) compared to simpler interactions.

Over time, you tend to develop similar gut microbiomes as those people you regularly spend time or share meals with outside the home, even if you're not related or living together. The previously mentioned US study published in 2019 found that siblings raised apart share fewer gut microbes than spouses do, highlighting that daily life together is a stronger force than genetics alone. While these studies used detailed genetic markers to track bacteria and showed clear links between social bonds and microbiome sharing, we can't definitively prove that the direct contact from social interaction (hugs, kisses, handholding) causes the transfer, as shared environments also play a role. But what they do reveal is that the people we share our daily lives with most intimately also become our closest 'microbiome buddies'.

WHY DO WE SHARE BUGS IN THE FIRST PLACE?

There is a fascinating theory that sharing our gut microbes might give groups an evolutionary edge. Think of it like this: when people in close-knit communities, such as families or tribes, share bacteria through everyday contact, they swap more than just germs. It has been suggested that they could, in fact, be exchanging beneficial microbes that boost the whole group's health.

These shared 'good' bacteria offer several potential advantages. First, they help train everyone's immune systems, making the group collectively better at fighting inflammation and allergies. Second, they act as a sort of defence squad, crowding out harmful pathogens and creating a kind of 'herd immunity' within the community. Third, they can unlock nutrients from shared foods, such as special bacteria or microbes that produce essential vitamins and helpful compounds. Essentially, sharing microbes could make the group healthier, hardier and better adapted to their environment and diet.

This raises an intriguing possibility: did some of our social behaviours, such as communal meals, shared childcare or kissing, actually evolve partly because they help spread these beneficial microbes? Observing animals certainly seems to support this; for instance, tightly bonded baboon groups or social macaques often share more health-promoting bacteria than less socially cohesive groups.

The big 'chicken-or-egg' question is: 'Did these social behaviours arise to spread helpful microbes? Or did the microbes simply hitch a ride on behaviours that evolved for other reasons, like cooperation or bonding?'

It's hard to prove microbes were the main driver. Furthermore, there seems to be a significant trade-off at play, because the same close contact that spreads helpful bacteria also makes it easier for dangerous pathogens to spread. For microbial sharing to truly be an evolutionary advantage, the health benefits gained from the 'good' bacteria must outweigh the risks posed by the 'bad' ones. Some scientists argue that the strongest evolutionary advantage of our social microbiomes might simply be an enhanced collective defence against invading pathogens.

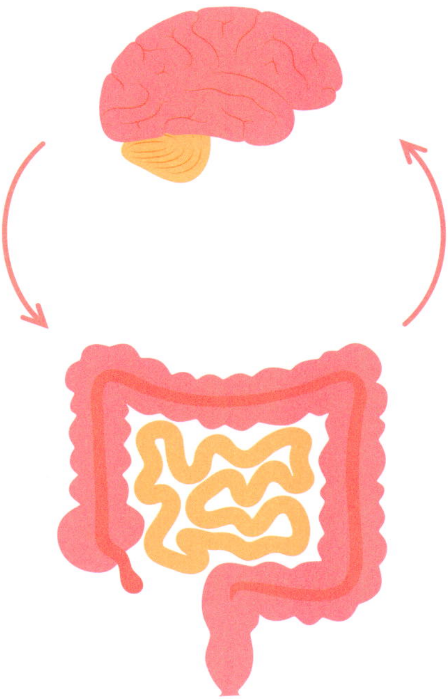

GUT-BRAIN AXIS

One of the most remarkable things these little bugs can do is influence your mood and the way you behave and think, quietly shaping how you interact with others. This is all thanks to a powerful and complex two-way communication network called the gut–brain axis that links your digestive system directly to your brain through nerves, hormones and immune signals.

Here's how it might work:

How your gut talks to your brain	What happens in your gut	How it helps your brain
Friendly acids from fibre	Gut bacteria ferment dietary fibre into small chain fatty acids (SCFA), such as butyrate, propionate and acetate.	These acids travel to the brain to help nerve cells grow (increasing levels of brain derived nerve factor BDNF) and connect (boosting learning and memory). They tell gut cells to make serotonin (the 'feel good' chemical). They calm brain inflammation (protecting brain cells).
Vagus nerve	The acids and other signals from bacteria activate the vagus nerve (the gut's direct line to the brain).	This nerve sends messages to brain areas that control emotions (amygdala) and stress (hypothalamus).
Stress control system	Gut microbes help calibrate your stress hormone system, the HPA axis.	Without normal gut bugs, animals have been shown to overreact to stress (high cortisol); adding microbes brings stress levels back to normal.
Brain barrier protection	Butyric acid helps seal up gaps in the 'brain shield' (blood–brain barrier).	This keeps harmful substances out of the brain, like a well-locked gate.
Brain's clean-up microglia	Gut microbes 'train' microglia (the brain's immune guards).	These microbes help manage inflammation and immune responses in the brain.

While the human evidence gathered so far largely shows associations, not definitive proof of cause and effect, intriguing patterns do exist that support a link between your gut and social brain. People with social anxiety or autism spectrum conditions often have gut bacterial communities in which alternative microbes thrive and work differently, even though the overall number of species isn't always lower. And early human studies suggest that specially formulated probiotics might subtly influence social decision-making, such as making people more willing to enforce fairness. There's even speculation that microbes could affect food cravings (steering us towards foods they thrive on) or partner selection through subtle scent cues, though this is largely theoretical.

The trillions of microbes living within you actively take part in a complex conversation with your brain, potentially influencing how you feel, handle stress and connect with others. Researchers are now working hard to tease apart exactly how much your microbial 'fingerprint' shapes your social world. So, although your gut microbes probably don't 'decide' who your friends are, by influencing your mood and physiology, they gently nudge you towards environments or behaviours where certain friendships can flourish. The goal? Avoiding social isolation and ensuring maximal opportunities for microbial exchange, because a diverse gut microbe population is a happy one. And a happy microbiome equals a healthy you.

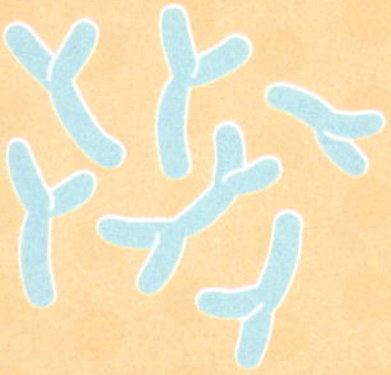

Bifidobacterium longum/bifidum

Lactococcus lactis

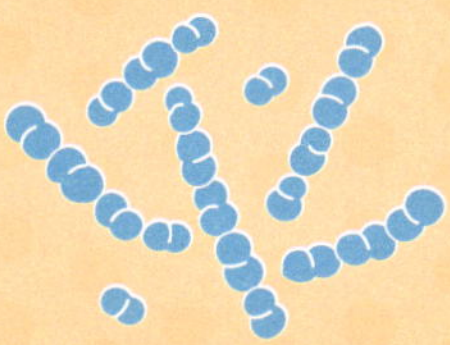

Streptococcus thermophilus

Enterococcus faecium

HOW TO EAT YOUR WAY TO A BETTER PERSONALITY

The connection between food and your psychological well-being has long been recognized. The term 'hangry' – that irritable feeling when you're hungry – first appeared in our vocabulary in the 1950s when the psychoanalytical journal *American Imago* used it while describing accidental wordplays. When you're hangry, it's the absence of food that triggers changes in your mood. But what about food itself? Can it actually alter how you feel about the person that you are?

Simply eating does seem to lift our spirits. Think about enjoying your favourite birthday meal or reaching for a sweet treat when you're feeling low. Those sugary snacks and desserts might temporarily boost your mood, but they elevate something else too: your blood glucose. But, as the saying goes, 'What goes up must come down.' When those sugar levels do crash, they bring fatigue, reduced alertness and irritability along with them. Talk about going full circle.

However, the question of whether food can affect your *personality*...well that's an entirely different question, because your personality is more enduring than your mood, which can fluctuate over hours or days. Your personality is the way you characteristically think, feel or behave due to neurobiological wiring and learned coping strategies that are shaped by your genes, early life experiences and long-term environmental factors, such as culture and family upbringing. Being happy, sad or irritable is very different to being introverted, open or conscientious.

Despite their differences, these aspects of your psychology are interconnected: your baseline personality shapes how prone you are to certain moods. For example, a highly neurotic person experiences negative moods more frequently. However, the moods you consistently experience over time can also reinforce or even gradually shift your personality traits. Given this relationship, could consistently eating mood-enhancing foods also influence personality traits over time?

In truth, the relationship between food and personality is very complex. While food choices and certain nutrients can influence mood and behaviour, there's no strong evidence that simply changing your diet will fundamentally alter your core personality traits. But diet and personality are linked in several ways, and some eating patterns may support better emotional stability and mental health.

Dietary pattern	Associated personality traits	Key findings[4]
Plant-based, fish, vegetables	Openness, conscientiousness, emotional stability	Linked to better mood and emotional regulation
High sugar, processed foods	Neuroticism, low conscientiousness	Linked to mood swings, emotional eating and lower emotional stability
Meat-heavy, high-calorie foods	Extraversion, neuroticism	Associated with impulsive and emotional eating

WHAT'S KNOWN:

- Diet can influence your mood and emotional well-being
- Personality traits predict food choices and eating behaviour

UNCLEAR:

- No strong evidence that specific foods or bacteria can directly cause lasting change in your core personality traits, such as openness or extraversion

[4] These findings draw on several studies published between 1998–2017 from Australia, Switzerland, the UK and the USA, and include large national and population surveys, cross-sectional and psychometric survey studies (psychometric surveys use validated psychological scales).

SUGAR AND ITS EFFECTS ON YOUR MENTAL HEALTH

When you eat lots of sugar, it floods your brain's reward circuits, mainly dopamine pathways, which amplifies your 'feel-good' response. Over time, your brain adapts to this by needing bigger sugar hits to help you get the same pleasure. This fuels those impulsive cravings and emotional eating episodes that all sugar lovers experience.

A high-sugar diet also impacts your body's stress management system, the HPA axis, which controls cortisol release in response to stressful events. High sugar disrupts this axis because, under normal conditions, a stressor, such as a tight deadline, triggers a quick burst of cortisol: the 'spike'. This helps you to focus and mobilize energy, and then settles back down when the challenge passes. When you consistently eat lots of sugar, this system eventually goes off-kilter. This results in two possible misfires:

1. Exaggerated spikes, where even a small stressor sends cortisol rocketing far higher than it should. You end up feeling jittery and anxious, and your body stays in high-alert mode, wearing out your system.
2. Flattened spikes, where your cortisol levels barely budge when a real stressor hits. You feel fatigued and unfocused, and your body can't mount a solid stress response.

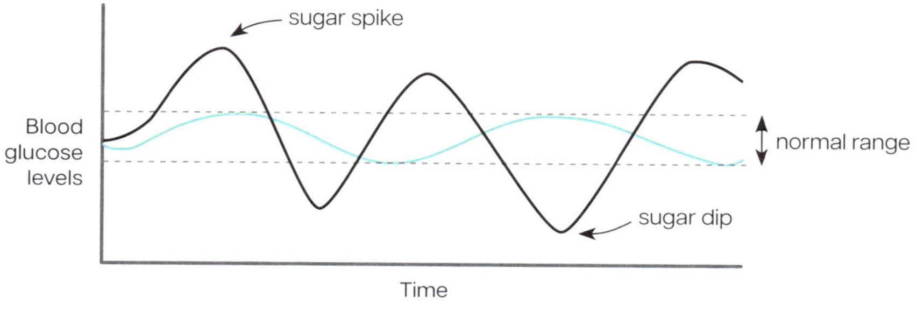

Large blood-sugar swings are harmful; keeping levels stable is healthier than repeated spikes and crashes. Those swings can distort mood and behaviour, while stability helps people feel more like their 'normal' self.

In both cases, your stress response loses its healthy up-and-down rhythm, which can leave you feeling either chronically wound up or chronically worn out. This habitual overload on sugar will ultimately cause your cells to become less sensitive to insulin and result in a condition called insulin resistance, which is linked to low-grade inflammation in the brain and body, elevating depression risk and potentially shifting your emotional baseline. So, this combination of hyper-reactive reward circuits, disrupted stress hormones and inflammation creates a perfect storm for:

- Increased impulsivity (acting without thinking)
- Emotional instability (rapid or exaggerated mood swings)
- Greater vulnerability to anxiety or depressive moods

If you consistently pursue a high-sugar diet, this will rewire key neural and hormonal systems and gradually alter your typical mood patterns and characteristic ways of behaving. But the type of sugar is important here. Processed, 'added' or refined sugars in sweet foods and beverages are more strongly linked to mood disturbances, depression and increased risk of common mental disorders. Natural sugars found in most whole fruits and vegetables are less likely to cause rapid blood-sugar spikes when they're consumed in their natural state *along with the fruit's fibre* and generally aren't associated with the same negative mood effects.

Blood sugar levels are just one way your diet can impact your mood. The influence gut microbes have, however, is on a totally different level.

ROLE OF THE GUT

Your gut and brain are in constant, sophisticated communication through the gut–brain axis. This dialogue happens via nerve signals (primarily the vagus nerve), hormones, immune molecules and key chemicals produced by your gut microbes. These bacteria actively produce, break down or influence neurotransmitters like serotonin, dopamine and GABA, which are critical for regulating mood, anxiety and motivation.

It's important to note that, while gut bacteria produce these neurotransmitters, these molecules don't directly travel to the brain. Instead, they exert their influence by supplying precursors and metabolites that shape the brain's own chemical environment. Changes in gut bacteria often accompany mood changes, but true cause and effect is shown when interventions such as specific probiotics or faecal transplants produce measurable changes in mood or behaviour. Animal experiments frequently demonstrate these causal links, whereas evidence in humans is weaker and needs more research to clarify timing and mechanisms.

PROBIOTIC STRAINS AND SOCIAL ANXIETY

You can influence your gut–brain system by taking psychobiotics, which are specific probiotic strains shown to support mental health. Strains such as *Bifidobacterium longum* and *Lactobacillus plantarum* have human-trial evidence for reducing anxiety; they likely work by modulating the vagus nerve, altering neurotransmitter production and regulating stress hormones like cortisol, which can improve emotional steadiness and social confidence. These strains appear in fermented foods such as kimchi, kefir and yogurt. Ordinary probiotics like *Lactobacillus acidophilus*, *Bifidobacterium lactis* and *Streptococcus thermophilus* may help general gut health but do not necessarily have proven direct effects on mood or behaviour.

Any direct effects of these psychobiotics on personality traits remains an open area of research. They probably don't rewire your personality, at least not in any proven way.

FEED YOU MOOD

Your brain depends on neurotransmitters to regulate mood, focus and feeling, and those chemicals are made from nutrients you get in food. Lacking the right nutrients reduces neurotransmitter production and can undermine emotional balance, while eating the right foods supplies your brain with the raw materials it needs to support good mental health. The table opposite provides some targeted food sources of neuroactive substances to help feed your brain.

Function	Key nutrients/precursors	Top food sources
Serotonin (mood, sleep, appetite)	Tryptophan (essential amino acid); B6; folate; iron; vitamin D	Eggs; dairy; soy; leafy greens; fortified cereals
Dopamine and noradrenaline (motivation, focus, reward)	Tyrosine (amino acid); B6; folate; copper; vitamin C	Lean meat; dairy; legumes; citrus fruits
Acetylcholine (learning, memory, attention)	Choline (essential nutrient); vitamin B5	Eggs; liver; soybeans; peanuts
GABA (gamma-aminobutyric acid) (relaxation, stress reduction)	Glutamate (neurotransmitter); vitamin B6	Protein-rich foods; meat; fish; legumes
Reducing neuroinflammation; supporting brain cells	Omega-3s: eicosapentaenoic acid (EPA) and docosahexaenoic acid (DHA)	Fatty fish (salmon, sardines); walnuts; flaxseeds
Neurotransmitter synthesis; cognition	B vitamin complex: B6; B9 (folate); B12	Whole grains; legumes; leafy greens; fortified breads
Neuroprotective, anti-inflammatory	Mediterranean dietary pattern: omega-3s; B vitamins; antioxidants; minerals	Fruits; vegetables; nuts; whole grains; olive oil; fish

However, it's worth noting that everyone reacts differently to dietary changes and its effects on mood vary depending on your particular genes, lifestyle, age, sex, health status and baseline gut microbiome diversity.

Gut–brain research is exciting but still has its limits. The hype about probiotics changing your personality runs ahead of the evidence, as most studies point to small boosts in mood, stress levels or emotional regulation, especially if you're already feeling low. Placebo effects are strong, and there's no solid proof they can reshape core traits like extraversion or conscientiousness. A more grounded takeaway would be accepting that personality tends to guide what and how we eat, and certain diets can, in turn, support mental wellbeing which, over time, can subtly influence how you think, feel and behave.

WHY CANCER RARELY SPREADS TO THE SPLEEN

Bowel cancer, one of the most common malignancies of the gut, can develop with few or no obvious symptoms for some time, so many people are not diagnosed until the cancer is at an advanced stage. When someone's told they have stage IV cancer, it usually means that the cancerous cells have spread from their primary site of origin to other parts of the body. The technical term for this process is 'metastasis'. For example, if someone originally diagnosed with colon cancer found out that it had spread (metastasized) to the liver, they are said to have 'metastatic colon cancer with liver involvement' or 'secondary liver cancer originating from the colon'.

This breaking free from the original site represents the most advanced stage in the cancer journey, as there is no stage V. A stark betrayal of the body's delicate homeostatic balance, cancer reveals its cruel nature through a relentless drive to colonize new organ territories, undermining the very order that keeps cells and tissues in harmony.

Amidst the vulnerable organ sites that are typically targeted, such as the liver, brain, bone and lungs, one organ stands remarkably resistant: the spleen. Why is the spleen so rarely colonized by these treacherous cells? What makes it an 'inhospitable fortress'? The answer may lie in the unique biological defences that it harbours against cancer's ultimate betrayal.

OESOPHAGEAL CANCER

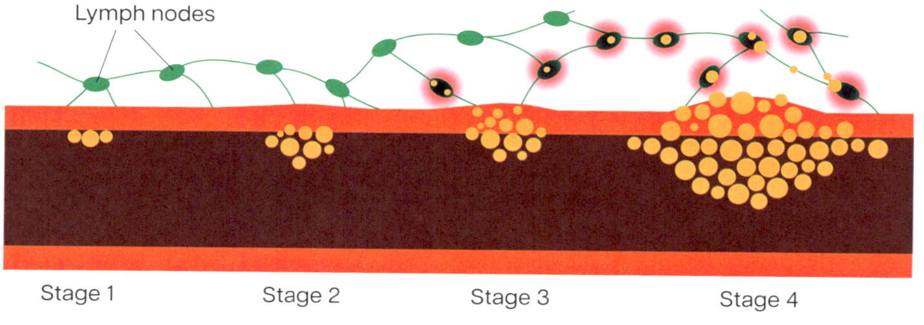

IT'S SO RARE

Splenic metastasis (cancers that spread to the spleen) is exceptionally uncommon			
<1% of all metastatic cases involve the spleen (large hospital study)	Very few per cent of all cancer patients develop splenic metastasis	0.6–7% autopsy studies show splenic involvement	Extremely rare isolated splenic metastasis cases

Common versus rare metastasis sites				
Liver Very common	Lungs Very common	Bones Very common	Brain Common	Spleen Exceptionally rare

Most common primary cancer sources when splenic metastasis does occur				
Lung cancer Hematogenous spread	Breast cancer Late-stage disease	Colorectal cancer More likely isolated	Ovarian cancer More likely isolated	Melanoma Immune evasion

Key clinical insights		
Context matters Most splenic metastases occur alongside widespread, multi-visceral disease rather than as isolated cases	Primary versus secondary Secondary (metastatic) splenic cancer is more common than primary splenic cancer, but both remain exceptionally rare	Biological mystery The spleen's anatomical and immunological features likely explain this rarity, but exceptions such as isolated splenic metastasis from melanoma and other rare cases of solid tumours that somehow overcome these barriers remain puzzling

The figures in this article are drawn mainly from retrospective series and case reports, reflecting the rarity of splenic metastases and the heavy reliance on autopsy and highly selected clinical cohorts in this area. As a result, the quoted percentages should be interpreted as approximate ranges rather than precise population incidences, since factors such as selection bias, variable detection methods and underreporting are likely to influence the observed rates.

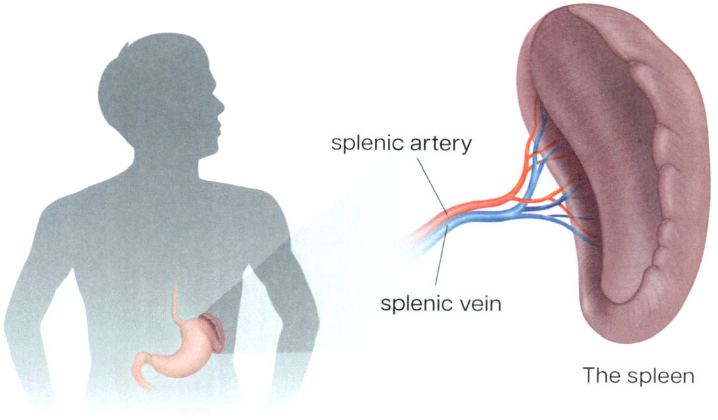

WHY IT'S SO RARE: THE SPLEEN'S DEFENCES

If we deconstruct the spleen's fortress, we'll find a veritable cache of weapons and armoury that are, in theory, at its disposal. This includes circulatory, immune surveillance, microenvironmental and biochemical, and mechanical factors that make it an inhospitable environment for any potential invaders. It is the spleen's unique architecture that creates the first line of defence. Unlike other organs with tightly woven capillary networks that cancer cells can easily infiltrate, the spleen features an 'open circulation' system. Blood flows freely through loose, spongy chambers called sinusoids, making it nearly impossible for tumour cells to stick to vessel walls and break through – the essential first step in establishing a new tumour colony. Adding to this challenge, the main artery feeding the spleen takes a sharp, almost 90-degree turn that acts like a mechanical filter, potentially blocking cancer cells from entering in the first place.

The spleen also cuts off a major invasion route that other organs can't avoid. While cancer commonly spreads through lymphatic vessels (the body's secondary circulation system), the spleen has no incoming lymphatic channels. In this way, it's like a castle with one less gate for enemies to breach.

But the spleen's most formidable defence may be its resident army. This organ houses an extraordinary concentration of immune warriors – natural killer cells, macrophages and specialized T-cells (cytotoxic T-cells) – that patrol the blood like sentries. For any rogue cancer cell, it's akin to walking into a lion's den. These cells don't just recognize and destroy invading cancer cells – they coordinate sophisticated attacks that can eliminate threats before they establish a foothold. These immune cells also enforce the spleen's day job as the body's recycling centre, which also works against cancer. As it breaks down millions of ageing red blood cells daily, it releases iron and haem – byproducts that prove toxic to many cancer cells through a process called ferroptosis, essentially rusting them to death. Such constant, aggressive filtering of abnormal cells naturally extends to cancer cells.

For any wandering cancer cells that do somehow manage to take root, the spleen creates a biochemical 'no-go zone' by releasing anti-tumour proteins (such as protease inhibitors and angiogenesis blockers) and unleashing reactive oxygen/nitrogen species that trigger cell death. Its iron-rich, acidic environment and voracious immune-cell metabolism also deprive rogue cells of the nutrients and pH balance they need to attach, grow blood vessels and survive, so metastases rarely establish there.

Lastly, the spleen's physical structure and function may repel invaders. Its muscular capsule contracts rhythmically, creating an environment of constant motion that could physically disrupt any cancer cells attempting to settle.

SEED AND SOIL

The 'seed and soil' hypothesis, first proposed by the English surgeon, Mr Stephen Paget in 1889, likens metastatic cancer cells to seeds that must find congenial soil (that is, the specific microenvironment of a distant organ) in order to grow into secondary tumours. According to Paget, circulating tumour cells (the seeds) will only colonize certain tissues (the soil) whose local conditions – such as available nutrients, the supportive framework around cells (extracellular matrix) and resident stromal cells that help maintain tissue stability – support their survival and outgrowth.

The spleen's tissue matrix appears to lack key adhesion molecules (for example, certain integrin ligands that act like a sort of molecular glue connecting a cell to its external world) and growth-promoting molecules called chemokines that many metastatic cells need to attach, invade and proliferate. Without these specific integrin-mediated binding sites or chemokine gradients to direct their homing and sustain their growth, cancer cells passing through the splenic microenvironment are less able to form new colonies, and this makes the spleen an inhospitable, 'infertile' soil for metastasis.

EXCEPTIONS

When cancer does reach the spleen, it unfortunately typically signals that the patient has an advanced, widespread disease. With most organs being infiltrated, it represents a grim milestone. Isolated splenic metastases (primary cancers that have only spread to the spleen) remain medical rarities. They typically originate from ovarian, colorectal or melanoma tumours that somehow breach the spleen's defences.

Perhaps most intriguingly, though, is the observation that some cancer patients develop new metastases after spleen removal, suggesting this organ might protect not just itself, but other parts of the body too. While direct evidence in humans remains limited, splenectomized patients show subtle increases in infection risk and possibly altered cancer progression patterns. These observations hint at the spleen's broader role as an immunity guardian: our quietly standing sentinel against cancer's spread throughout our bodies.

WHAT WE CAN LEARN FROM THE SPLEEN

The spleen's remarkable resistance to metastasis has captured the attention of cancer researchers worldwide, who see it as a possible biological blueprint for new treatments. Scientists are exploring whether they can harness the spleen's secrets, perhaps by boosting the body's natural killer cells that patrol this organ so effectively, or by developing drugs that recreate the spleen's hostile microenvironment in other tissues. Another potential explorative avenue could involve targeting the adhesion pathways that the spleen naturally lacks, potentially preventing cancer cells from establishing footholds elsewhere in the body.

While these approaches remain largely theoretical and confined to laboratory studies, the spleen's unique properties offer invaluable insights into how organ-specific environments influence cancer's ability to spread. Understanding why this organ stands so firmly against metastasis could revolutionize how we think about cancer treatment and perhaps one day potentially transform the body's other tissues into equally inhospitable territories for tumour cells.

HOW CANCER BETRAYS US

In the battle between order and chaos, cancer emerges as a betrayal from within. Normal cells, once governed by the body's intricate checks and balances, begin to defy protocol. This transformation is driven by changes in a cell's DNA sequence (genome) and the way the code is read and used (epigenome) to dismantle the regulatory systems responsible for controlling cell growth, division and death.

Mutations in key genetic players (oncogenes and tumour suppressor genes) enable rogue cells to bypass signals that would ordinarily slow their replication or trigger their demise through apoptosis, the body's programmed cell-death system.

Uncontrolled cell division and the evasion of apoptosis are hallmarks of cancer. To survive, cancer cells hijack powerful molecular pathways. The intricate control systems (PI3K/Akt/mTOR and Wnt/β-catenin signalling pathways) present in all cells that normally dictate how they grow, survive and manage energy become allies in the cancer's campaign to ignore the body's restrictions.

FROM SURVIVAL MECHANISMS TO TOOLS OF SELF-DESTRUCTION

In its manic obsession for survival, cancer plays a dangerous game: it repurposes the very tools life relies on. Cancer doesn't just defy death; it turns survival into a self-destructive weapon. Cells that should perish somehow adapt and persist, transforming physiological lifelines into mechanisms of destruction.

In healthy organs, detachment from the surrounding tissue triggers a form of cell death known as 'anoikis'. But cancer cells rewrite the rulebook. They resist anoikis, drifting through the bloodstream and settling in distant organs as seedlings of metastasis.

Tumour cells, especially cancer stem cells, enhance DNA repair machinery and activate drug-efflux pumps: molecular gatekeepers that eject therapeutic agents. In doing so, they deflect any chemotherapy and radiation meant to destroy them. They have now succeeded in achieving 'chemo-resistant' status.

To endure hostile environments, cancer cells rewire their metabolism. This reprogramming allows them to flourish under stress, consuming and converting resources in ways normal cells simply can't and feeding their unchecked growth.

FANCY A SENEGALESE TONIGHT, DARLING?

As takeout options go, Chinese, Indian or Mexican cuisines are commonplace staples in the Western diet, introduced via diaspora communities from their regions of origin. Foods from the African continent, in particular West Africa, aren't as common, although they're gaining popularity in many Western countries. Take jollof rice, for instance – this vibrant, tomato-based dish has become increasingly popular in countries like the USA, Canada and beyond.

Originating from the Wolof people of present-day Senegal, The Gambia and Mauritania, jollof rice has been embraced by every West African nation, particularly Ghana and Nigeria, who over the centuries have adapted it to their local tastes and ingredients. Jollof is so central to West African culture that locals say, 'A party without jollof is just a meeting.'

But jollof rice isn't the only food from this region that is worth trying. In fact, from a health perspective, other dishes are far superior. And, given the well-documented health risks associated with typical Western diets, research suggests that the more traditional foods eaten in many African countries might hold the keys to better health in ways that we're only beginning to understand.

TRADITIONAL AFRICAN DIET

Traditional cuisine in Africa represents one of the world's most diverse and nutritionally sophisticated food systems. From the Chagga communities of Mount Kilimanjaro to the savannas of West Africa, traditional diets across the continent share a remarkable foundation. Wild-harvested foods, fresh vegetables and hearty whole grains such as sorghum, millet, maize and teff form the core of these culinary traditions. And these ingredients are often paired with protein-rich legumes, such as black-eye peas, cowpeas and bambara nuts, creating nourishing and diverse meals unique to each country.

What makes these diets truly distinctive is their emphasis on minimal processing and fermentation. *Kenkey*, a fermented maize dough that's a daily staple in Ghana; *ogi*, a fermented porridge eaten at breakfast in Nigeria; *injera*, a fermented teff flatbread from Ethiopia; and *mbege*, the traditional Chagga banana-millet brew drunk in Tanzania's highlands, all typify traditional meals containing beneficial microbes that transform simple ingredients into complex, nutritious foods. Old-school preparation methods, such as boiling, steaming, fermenting and drying, preserve nutrients while developing deep, satisfying flavours without relying on added sugars or excessive fats.

Starchy roots such as cassava and yam are often eaten with stews made from fresh onions, peppers and tomatoes, or green leafy vegetables, while wild fruits foraged from local landscapes replace processed desserts. Alongside conventional sources of protein such as poultry, fish and cattle, some regions also incorporate bushmeat and edible insects into their diets.

This approach delivers exceptional nutrition: high fibre, abundant polyphenols (powerful antioxidant and anti-inflammatory compounds), healthy omega-3 fatty acids and minimal saturated fat or added sugars. Unlike Western diets that are typically heavy in processed foods, refined grains, red and processed meats, sugar, salt and saturated fats, with low intake of fibre and plant-based foods, these traditional patterns offer a masterclass in eating for both pleasure and health.

TEMBA STUDY

In 2021, a groundbreaking experiment took place that challenged what we thought about diet and disease. Dr Quirijn de Mast and Dr Godfrey Temba's team assembled 77 healthy young men from Tanzania's rural and urban communities for a study that would see how quickly and how much a person's body reacts when they swap a traditional, plant-rich African diet for a modern, Western-style one, and vice versa. The question was: would these short-term diet swaps boost or impair health?

The setup was simple: rural men accustomed to traditional African-heritage diets (rich in fermented foods, whole grains and vegetables) were asked to switch to Western diets heavy with processed foods. Their urban counterparts, who were already consuming Western-style meals, were guided back to their ancestral eating patterns. The study also included a third group that consumed the fermented beverage *mbege* on a daily basis. Within just two weeks, the researchers witnessed transformations, and their findings were eventually published in the journal *Nature Medicine* in 2025.

The body's swift response

When rural participants abandoned their traditional diets for Western food, their bodies launched what could only be described as a biological alarm system. The Western diet triggered the production of 26 inflammatory proteins and molecules associated with heart disease and chronic inflammation. And their immune systems, once finely tuned, began showing signs of what scientists call 'immune refractoriness' – a dulling of the body's ability to fight infections.

But perhaps most striking was what happened when the process was reversed. Urban dwellers returning to their heritage diets saw a remarkable cleanup operation unfold within their bloodstreams. Twenty-eight inflammatory proteins decreased, including key players such as IL-6 and various CXCL chemokines, the molecular messengers of inflammation that fuel chronic disease.

Fermentation

The third group's experience with mbege revealed something extraordinary about fermentation's power. This humble banana beverage didn't just provide nutrition; it appeared to reprogramme immune function at the cellular level. Participants showed increased production of IL-10, the body's natural anti-inflammatory mediator, while their immune systems developed what researchers termed 'immune training', a heightened ability to respond to threats such as Candida (fungus) infections.

The metabolic changes were equally dramatic. Mbege consumption flooded the bloodstream with compounds such as hippuric acid and valerolactones, which are breakdown products of polyphenols that act as potent antioxidants. These changes represented the gut microbiome actively processing and transforming plant compounds into health-promoting molecules.

Molecular biology

Using cutting-edge multi-omics analysis that looks at many layers of biology at once – genes, proteins, chemicals and more, the researchers essentially performed molecular archaeology on the participants' blood and discovered that dietary interventions explained an astonishing 40 per cent of the variance in immune responses and metabolic profiles. Such a degree of influence rivals genetic factors.

The Western diet left its fingerprints everywhere: elevated urea cycle metabolites – the intermediate chemicals produced as the body converts toxic ammonia into urea for excretion (suggesting kidney stress), disrupted glucose homeostasis pathways (the body's systems that normally keep blood sugars steady) and a shift toward inflammatory omega-6 fatty acids at the expense of protective omega-3s. Meanwhile, the restoration of the heritage diet brought increases in compounds such as resveratrol glycosides (plant antioxidant derivatives) and dramatic reductions in uremic toxins, waste products that the kidneys struggle to clear and that contribute to cardiovascular disease.

Persistent change

What was most remarkable was the study's final revelation: many of these changes from adopting the heritage diet persisted four weeks after the interventions ended. The body seemed to remember its dietary lessons, maintaining lower inflammatory protein levels and improved metabolic profiles long after participants had returned to their usual eating patterns.

This finding challenges the conventional wisdom that dietary benefits disappear quickly. Instead, it suggests that even brief exposures to traditional foods can trigger lasting biological reprogramming – a kind of metabolic memory that could have profound implications for preventing chronic disease.

Despite its novel insights into dietary 'memory', the trial was not without its limitations, and these would need to be addressed to provide absolute certainty that the effects seen were not the result of random chance. The fact that the trial enrolled only young, healthy men from a single region in northern Tanzania limits how generalizable these findings are to women, older adults or those with chronic disease. Interventions were also brief (two weeks for diet swaps and one week for the fermented beverage, with only a single four-week follow-up), leaving us uncertain about any long-term effects.

The researchers also used an open-label design (where both the researchers and participants know which treatment each participant is receiving) and relied on self-reported recalls of their baseline diet outside provided meals for eligibility. These factors, combined with potential bias from the financial motivation of receiving free meals, may have influenced the study's outcomes and sample composition. Finally, the modest sample size and focus on indirect molecular signals rather than actual improvements in the participant's health constrain the study's real translational impact. The Temba study provides a roadmap for understanding how traditional dietary wisdom and practices might combat the rising tide of non-communicable diseases that currently plague the West, and that are now sweeping across 'modernizing' nations in Africa and beyond. As urbanization drives populations away from their ancestral diets, this significant research suggests that the path back to health might be found not in modern pharmaceuticals, but in the fermented beverages and whole foods that sustained human health for millennia.

Topic	What's happening	Why it matters	What you can do
Dietary transition	Urbanization shifts diets from traditional whole foods to processed, Westernized options	Leads to a double burden: undernutrition and rising chronic diseases	Choose local staples (millet, greens) over packaged foods Prepare traditional fermented drinks and meals
Health impacts	City dwellers have more inflammation and non-communicable diseases than rural peers	Chronic inflammation precedes heart disease and diabetes, reducing life quality	Eat anti-inflammatory foods (whole grains, legumes, fermented foods) Screen regularly for BP, sugar, inflammation
Barriers to traditional diets	City growth, high cost of heritage crops, aggressive marketing of cheap processed foods and loss of cooking knowledge	People default to unhealthy processed foods, increasing health risks and eroding cultural identity	Start/join a community garden for local crops Share family recipes in groups
Community and policy solutions	Evidence supports farm subsidies, nutrition education, and sugar taxes	Policy can make healthy traditional foods affordable and valued	Advocate for subsidies for local grains Push for cooking in school lessons Support 'junk food' tax campaigns
Research and innovation	Indigenous foods are understudied; few trials show their benefits	Lack of data hinders investments/efforts to protect and promote these diets	Collaborate with universities/NGOs to document local food practices Champion community health trials
Reclaiming dietary sovereignty	Heritage diets are anti-inflammatory, nutritious and sustainable	Reclaiming ancestral foodways improves health, food security and cultural identity	Host a traditional meal potluck Try one traditional recipe weekly Shop at local markets for indigenous produce

WHY ARE PUTRID, SMELLY THINGS ALWAYS DEPICTED AS BEING GREEN?

The churning, queasy feeling in your gut when something disgusts you is so primal that we've even designated it a colour, and painted it green in our imaginations. In cartoons, animations, books and movies, scary, grumpy and cantankerous characters such as ogres, aliens and even sometimes monsters are more often than not green. Whenever someone farts, has bad breath, or opens toxic or decaying waste barrels, you'll notice that the colour green is nearly always used. But why green, when most smelly odours are colourless, and disgusting, rotting or scary things come in all colours? Is there any biological basis to this, or is it purely the result of a random decision made by a Hollywood executive to use the colour and it somehow stuck? Time to journey from biology to motion-picture boardrooms to find out if there's any science behind why the green-equals-gross trope exists.

WHAT NATURE TEACHES

In nature, green is often associated with life, freshness, balance, growth, renewal, fertility – you name it, it's all 'kumbaya', peace and harmony. So, there's a bit of a... no, actually, there's a massive disconnect between these associations and green being used to represent unpleasant odours and unsavoury characters.

The green we see in nature comes from the green pigment chlorophyll in plants. But when plants die, the colour green doesn't magically intensify as the green-equals-gross trope suggests. Instead, it breaks down into colourless compounds through a precise biological recycling process. So, the next time you see a decaying plant or leaf, know that what you're actually seeing are the underlying yellows and oranges from carotenoids (fat-soluble antioxidants) which were previously masked by the dominant green colour of chlorophyll that eventually fade to brown.

As autumn arrives, chlorophyll pigment breaks down, green fades and the vibrant autumnal colours emerge.

Now I'm not suggesting that green isn't associated with anything repugnant or alarming in nature. In fact, quite a few things spring to mind (see overleaf).

Scenario	Examples	What makes it green	When and where you'd see it
Bodily fluids and excretions	Green pus in bacterial infections	Myeloperoxidase (an enzyme found in neutrophils, a type of white blood cell) oxidizes compounds, producing green-hued pigments	In an infected cut or abscess, you might see thick, yellow or green pus, signalling a neutrophil-driven immune response
	Healing bruises turning greenish	Biliverdin, a green pigment produced from oxidized bilirubin (a yellow breakdown product of haem, the iron-containing molecule in red blood cells)	5–10 days after a bruise forms, the confined blood's haemoglobin is broken down to green biliverdin before fading
	Bile		If you throw up on an empty stomach, the vomit might contain bile, appearing greenish
Decomposing organic matter	Algal slime on wet logs or pools Green slime mould (e.g. *Fuligo* species)	Chlorophyll (from algae) or pigmented spores in slime moulds	After rain, look for green sheens on wood, rocks or stagnant water as evidence of algae On mulch or forest floors, yellow green 'dog vomit' slime mould patches appear
Animal defences and secretions	Tree-frog skin secretions	Biliverdin or carotenoid-related pigments in amphibian mucus	Touching brightly coloured frogs might leave a green tacky residue to warn predators of toxins
	Hagfish slime	Protein mucins plus trace pigments in hagfish slime	Disturbed hagfish spray green 'slime' to clog the gills of attackers
Plants and fungi	Unripe green tomatoes (solanine)	Chlorophyll in the unripe fruit skin	Green, firm tomatoes are rich in chlorophyll, and in solanine, which can cause nausea if eaten raw
	Death cap mushroom (*Amanita phalloides*)	Fungal pigments	The iconic green cap of the death cap mushroom contains toxic fungal metabolites – avoid any wild greenish mushroom[5]
Pathogenic growths	Bread mould (*Penicillium, Aspergillus*)	Pigments present in conidia (mould spores). Some moulds also produce mycotoxins within these spores.	Fuzzy green patches on forgotten bread signal mould toxins – discard immediately
	Pseudomonas aeruginosa on wounds	Pyocyanin (a blue-green bacterial pigment)	A green-tinged infection in burns or catheters often points to *P. aeruginosa*, a dangerous opportunist

[5] Warning: Never consume wild mushrooms without absolute, expert identification. Misidentification can lead to severe poisoning, organ failure and death.

So, while in the wild, the colour green's association with health and vitality makes its use for grotesque imagery seem contradictory, the connection becomes clearer when we consider the many repugnant green things that also exist in nature. When green does appear in decay, it comes from specific culprits, such as the blue-green patches on mouldy bread that are caused by microbes such as *Penicillium* or *Trichoderma* producing spores. It's not the process of decay itself that's green – it's the microbes involved in that process.

And when it comes to danger signals in nature, you won't find toxic animals relying solely on green warnings. Why? Because green camouflages far too well in leafy environments. Instead, evolution seems to favour high-contrast alerts. For example, poison dart frogs use metallic green *alongside* bold blacks or yellows, and wasps, bees and hornets flash bright yellow stripes. In reality, it's red, yellow, black and white that nature uses as its universal 'danger' palette, because they 'pop' out against greenery.

WHY DO WE ASSOCIATE GREEN WITH ROT AND DISGUST?

Blame your brain's wiring. Nature hardwired you to react strongly to signs of decay, and colour plays a surprising role by triggering an ancient 'disgust reflex' designed to keep you safe. Think of it as your brain's early warning system: long before you touch or smell something dangerous, visual cues can make you step back. Disgust evolved as a lifesaver. Studies show that our ancestors who avoided contaminated materials, thanks to visceral reactions to colours, textures or smells, survived and reproduced more successfully. That wrinkled nose and narrowed gaze you make when something looks 'off'? That's not just you; it's a universal human response, fine-tuned over millennia to protect your eyes, nose and mouth from infection. So, when you recoil from spoiled, smelly food, it's your body's way of saying, 'This could harm you.'

But gross smells don't just make you recoil; according to recent research they can also trigger real immune reactions in your mouth by increasing a molecule called TNFα in your saliva that is associated with helping fight infections.

CULTURAL REASONS

Across history, different societies have linked the colour green to poison, illness and even the supernatural. But those connections weren't always rooted in reality. Historically, brown and black were more closely tied to decay, death and putrefaction, aligning better with the colours of decomposition. For instance, medieval and Renaissance art typically depicted death and decay in dark brown and black tones rather than green.

So, how did green become 'sickly'? From the 1500s to the early 1900s, many doctors diagnosed young women with chlorosis, claiming it turned their skin pale green. In reality, the condition was simply anaemia, and the patients weren't green at all. The colour was more symbolic. Culture had decided that green meant unwell long before science proved otherwise.

Sometimes, however, green really was dangerous. In the 18th and 19th centuries, vivid green dyes like 'Paris Green' and 'Scheele's Green' contained arsenic. These dyes were used to colour wallpaper, clothing and even children's toys that slowly poisoned people. The Victorians called it the 'arsenic waltz' – a grim nickname for deaths caused by these fashionable greens. So, while nature doesn't necessarily make green things poisonous, history has given us plenty of reasons to distrust the colour.

But do humans naturally dislike it? Not exactly. Studies show infants tend to prefer blues and reds; green just doesn't grab their attention. But as kids grow up, they absorb cultural clues, and by school age those connections stick. So, while no one is born with an aversion to green, society also wires us to link it with disgust.

A CONVENTION OF ANIMATION STORYTELLING

Animation supercharged the connection between the colour green and digust. Imagine you're an animator, and you have seconds to tell the audience something smells terrible or is rotting. Bright green pops against human skin tones (making stink lines or sickly characters unmissable), it echoes mould and slime, and it stands out on screen. Whether or not it's biologically accurate doesn't matter; it helps show 'disgusting' fast.

Of course, animation isn't real life. Just as characters stretch like rubber, and eyes pop out of heads, colours get amplified. Rotting food becomes glowing green, and bad smells take on floating green 'stink' lines. This shorthand, however, is largely a Western invention. Japanese anime often uses totally different colour codes – brown or cloudy white, as seen in the 'Stink Spirit' from Studio Ghibli's *Spirited Away*.

This cartoon trick didn't stay on screen. Kids' slime toys? Almost always neon green. The nauseated emoji? Also green. What began as a cultural and artistic shortcut has become a global visual code. Biology might have sparked some of our aversion to green things such as mould, but it was culture, and especially animation, that made the colour an enduring symbol of 'yuck'.

WHY DO I GET BUTTERFLIES IN MY STOMACH?

That familiar phrase 'butterflies in my stomach' has as a surprisingly specific origin. The American author Florence Converse is attributed as being the first person to use this metaphor when she wrote in her 1908 novel *The House of Prayer* about a man's travel-induced melancholy, which felt to him like having a butterfly in his stomach. Her original use of the phrase differed markedly from today's meaning. She employed the singular 'butterfly' to capture her character's melancholy, describing a wistful sadness, rather than the anxious fluttering sensation we associate with it now. The phrase evolved considerably over the decades, and by the 1930s, the plural form 'butterflies' emerged in popular culture and gained further traction during the Second World War, when paratroopers adopted the shortened 'butterflies' as military jargon for pre-jump nerves.

Through this cultural evolution, Converse's rather poetic metaphor for sadness somehow transformed into our modern expression for anxiety and excitement. The phrase probably stuck because it perfectly captures the distinctive fluttery sensation in the abdomen that accompanies intense emotions – be it anticipation before a first date, nerves during a job interview or that all too familiar moment when your name's just been called out to speak in front of lots of people.

IT'S A WEIRD FEELING

That familiar fluttery feeling in your stomach is a genuine physiological response that actually has a fascinating biological basis, though the complete picture is more nuanced than it might first appear. People describe the sensation in somewhat consistent ways: trembling or fluttering in the abdomen, waves of nausea, warmth in the stomach's pit, churning sensations and the uncanny feeling of actual butterflies moving inside. But what's causing all these sensations?

When your brain detects an emotional trigger, it instantly activates the sympathetic nervous system, your body's ancient fight-or-flight response. This releases stress hormones such as adrenaline and noradrenaline, which trigger a complex redistribution of blood flow throughout your body. While your blood rushes towards muscles needed for quick action, circulation to your digestive system becomes reduced and erratic. However, the effects on your digestive circulation are more complex and variable than a simple 'redirection away' from the stomach. Factors like the type of stress, for example, can determine the exact response. Physical exertion, or psychological stress can trigger different nerve and hormone signals that change where blood goes, including to or from the stomach and intestines. Individual differences such as feeding state, fitness level, sex hormones, age and autonomic nerve function can also shape how strongly gut blood vessels constrict. So, within your digestive system, these stress responses can alter blood flow and trigger changes in the way your stomach contracts. The millions of sensory nerves lining your gut detect these shifts, potentially contributing to that distinctive fluttery sensation.

Yet the story doesn't end with pure physiology. Research suggests that how you perceive these sensations may be significantly shaped by psychological expectations, cultural learning and past experiences. The very act of anticipating 'butterflies' might influence how intensely you feel them. This highlights the fascinating complexity of the gut–brain connection.

YOUR BODY'S ANCIENT ALARM SYSTEM

Imagine you're walking through tall grass when you hear a rustling sound behind you. In that instant, your body doesn't know if it's wind or a predator, but it prepares for the worst. This is your fight-or-flight response in action, an ancient alarm system that has kept humans alive for millennia. The moment your brain detects potential danger, it triggers a cascade of survival mechanisms. Your hypothalamus – a small but mighty region at the brain's base – kicks your stress systems into gear, prompting your adrenal glands to flood your bloodstream with adrenaline and cortisol. These chemical messengers instantly shift your nervous system from rest and digest to fight or flight, preparing your body for immediate action.

MAIN EFFECTS OF FIGHT-OR-FLIGHT RESPONSE

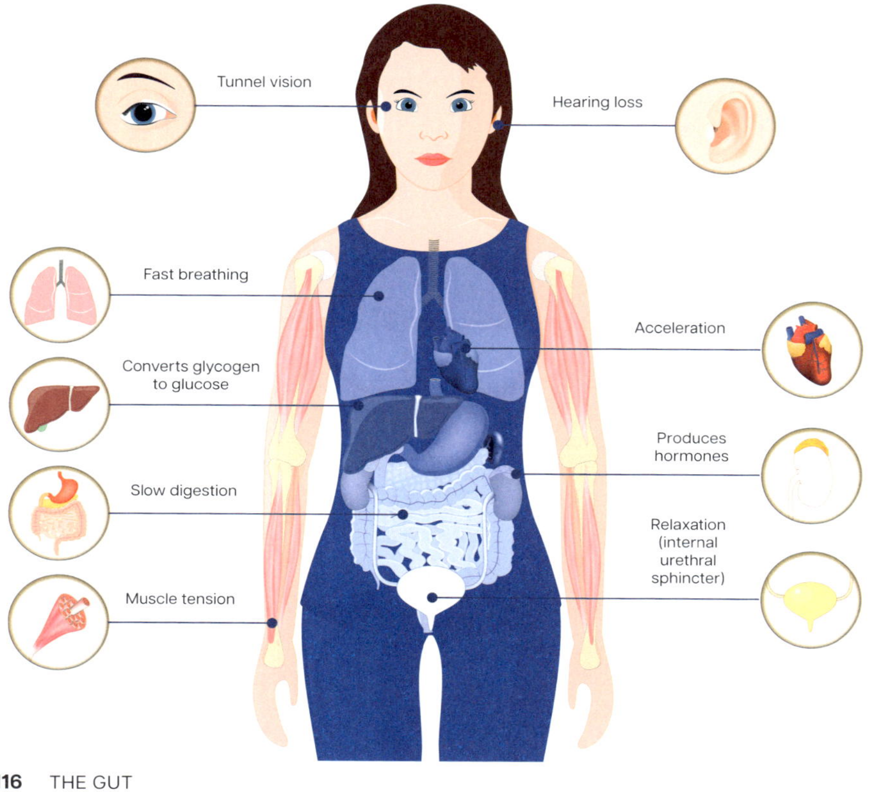

What happens next is a remarkable feat of biological engineering: your cardiovascular system becomes a tactical command centre, redirecting blood flow like a wartime supply chain. Blood rushes away from your digestive organs (because digestion can wait when you're potentially running for your life) and floods toward your skeletal muscles, heart and brain. Your muscles receive the oxygen and nutrients they need for explosive movement, while your brain gets enhanced blood flow to sharpen focus and quicken reaction times. (Science disclaimer: the specific effects on digestive motility and the exact patterns of blood-flow redistribution can vary and are not consistently observed across all types of stress or studies.)

This response evolved because it provided our ancestors with a crucial survival advantage. Those who could rapidly mobilize their body's resources were more likely to escape predators or defend themselves successfully. The system activates within seconds, demonstrating millions of years of evolutionary fine-tuning. While modern stressors rarely require us to outrun predators, this ancient alarm system continues to serve us, preparing our bodies to meet whatever challenges lie ahead.

THE NEURAL CONNECTION

Think of your vagus nerve as the information superhighway connecting your brain to your gut, a critical communication line that carries messages in both directions. This longest cranial nerve acts as the primary coordinator of your rest-and-digest functions, controlling everything from stomach acid production to intestinal movement. Remarkably, up to 80 per cent of its fibres carry signals from your gut to your brain, making your digestive system more of a storyteller than a silent listener.

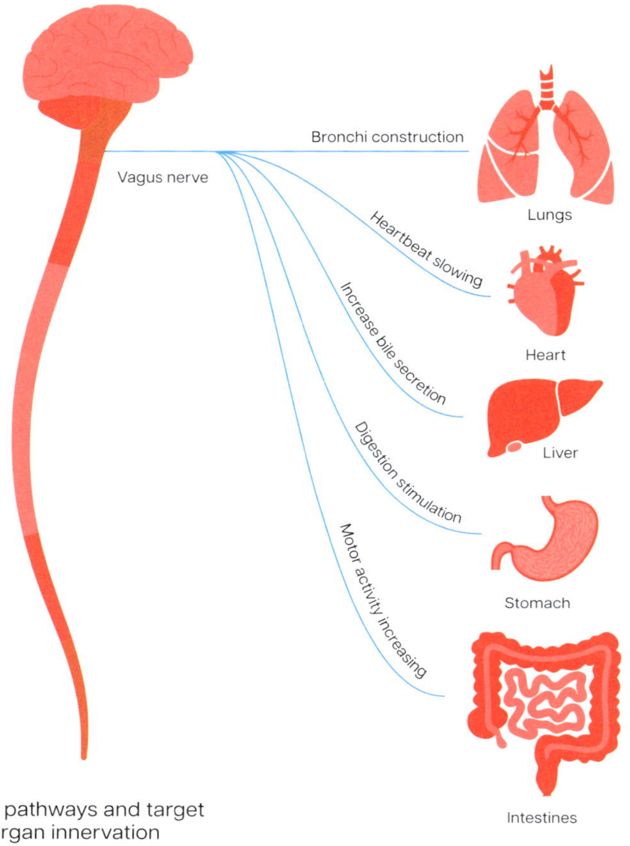

Vagal pathways and target organ innervation

When stress strikes, this sophisticated communication network experiences dramatic disruption. Your stomach muscles, which normally contract in coordinated waves to move food along, suddenly freeze up like a deer in headlights. The upper digestive tract essentially hits the brakes – gastric emptying slows, muscle contractions weaken and digestive enzymes reduce their output. Meanwhile, paradoxically, your lower digestive tract becomes hyperactive, explaining why in some people stress can simultaneously cause both stomach knots and urgent bathroom visits. This seemingly chaotic response is orchestrated by a stress hormone called corticotropin-releasing factor, which acts on different receptor types throughout your digestive system.

But there's another player in this story: your enteric nervous system, often called the 'second brain'. This network of hundreds of millions of neurons embedded in your gut wall contains more nerve cells than your entire spinal cord and can function independently of your brain, although your central nervous system still has a major say on how it behaves by tightly regulating its activity.

Your enteric nervous system produces the same neurotransmitters as your brain, including serotonin and dopamine, among others. It serves as the sensory mechanism that detects blood-flow changes during stress. When your cardiovascular system redirects blood away from digestion, these gut neurons are the ones responsible for registering the change and contribute to that distinctive fluttery sensation, creating the perfect biological storm we recognize as butterflies in the stomach.

ARE THERE DIFFERENT TYPES OF BUTTERFLIES?

The butterflies phenomenon occurs with both positive and negative emotions. Whether you're falling in love or facing a scary situation, both excitement and anxiety activate the same sympathetic nervous system pathways, triggering identical blood-flow redistribution patterns that create the fluttery sensation. However, the experiences differ in subtle ways: positive emotions typically produce shorter-lived butterflies that resolve quickly once the moment passes, while anxiety-related responses tend to linger longer and may be accompanied by additional stress symptoms such as sweating or nausea. People's sensitivity varies a lot, and feeling stronger 'butterflies' is just one end of that spectrum. It's not a flaw, just a natural difference in how our nervous systems react to emotional moments.

THE PREFRONTAL CORTEX

COGNITION, AMBITION AND BURNOUT

4.

WHAT'S THE BEST WAY TO STUDY ACCORDING TO NEUROSCIENCE?

Two students spend identical hours studying for an exam. One sits hunched over a laptop in a cramped, low-ceilinged dorm room with music streaming through their earbuds. The other works at a standing desk in a library's lofty reading room, eyes focused slightly upward, surrounded by silence. According to neuroscience research, the second student isn't simply more likely to absorb information. Rather, their brain is wired for enhanced creativity and deeper focus. And the difference? Well, it lies not in any additional talent or effort by student No. 2, but in their understanding of how the brain actually learns.

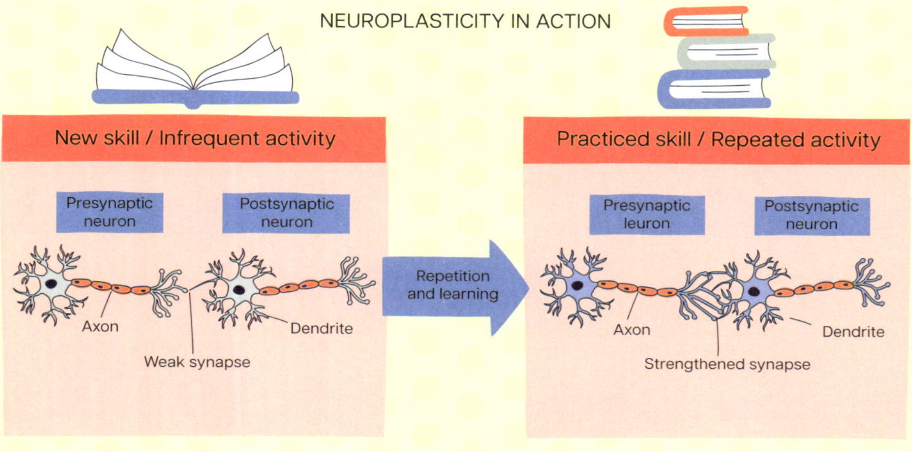

HOW YOUR BRAIN ACTUALLY LEARNS

The foundation of all learning lies in neuroplasticity, the nervous system's ability to change in response to experience. This process involves strengthening existing connections between neurons through 'long-term potentiation' – often summarized as 'neurons that fire together, wire together' – while simultaneously weakening unnecessary connections through a process called 'long-term depression' (a technical term not associated with mental health). The brain also forms entirely new connections between existing neurons, creating the neural highways that encode new knowledge.

The learning process is believed to unfold in two distinct phases. During focused attention, the brain signals that change is needed. The strain and agitation experienced when you consciously force your brain to pay attention triggers the release of neuromodulators such as adrenaline, which cue neural circuits for later modification. However, the actual rewiring and consolidation of memories occurs predominantly during deep sleep and rest states, when the hippocampus replays experiences and integrates new knowledge into your brain's cortex.

This dual-phase system explains why cramming often fails. The brain requires both intense focus to initiate change and adequate rest to complete the rewiring process. Most students who optimize both phases see dramatically better results than those who ignore this fundamental biological rhythm.

Effective studying isn't simply about time management or willpower. It requires three key elements:

- optimizing your physical environment,
- leveraging your brain's natural processes,
- strategically focusing your visual and cognitive systems.

OPTIMIZING YOUR PHYSICAL ENVIRONMENT

Recent research has uncovered surprising connections between physical space and cognitive performance. Joan Meyers-Levy and Rui Zhu's groundbreaking 2007 study published in the *Journal of Consumer Research* established what they termed the 'cathedral effect', demonstrating that ceiling height directly influences thinking patterns. According to their research, high-ceiling environments encourage expansive, creative problem-solving by allowing the brain to scale its functions with perceived space. Conversely, lower ceilings are more conducive to establishing the necessary focus for detailed, analytical work. Their work revealed that high ceilings create a sense of freedom that leads to broader thinking, while low ceilings trigger more detailed evaluations of specific items, a principle that traces back to Edward T Hall's observations in the 1960s about people feeling liberated in cathedrals versus enclosed in chapels. And it's not just a psychological thing – our brains are actually wired to align with the space around us. Over time, the nervous system has become pretty good at switching gears, shifting between the wide-open feel of outdoor environments and the more enclosed feel of indoor spaces. So the way we think can literally change depending on what kind of space we're in.

Visual attention drives cognitive focus in ways most students never consider. While research hasn't confirmed that a larger 'visual window' directly boosts learning, optimizing visual environments does enhance sustained attention. By reducing distractions and improving screen ergonomics we minimize physical strain and cognitive load. This matters because the brain relies on integrated sensory maps that coordinate vision, hearing and movement, making thoughtful environmental design essential for effective learning.

The strategic use of 'blinkers' that minimize visual distractions through physical barriers or environmental design represents another powerful tool. When surrounded by visual clutter or competing stimuli, your brain shifts to broad scanning (distributed processing) rather than narrow, focused attention. Highly effective students instinctively study alone in minimalist environments, putting phones away and eliminating other distractions. This creates what neuroscientists like to call 'neurosemantic coherence', where relevant neural networks activate strongly, while unrelated ones are inhibited. Or, in simpler terms, an intense focus where neural resources primarily align themselves to a single task or problem, minimizing distractions. This concept, often used in the context of deep work, is presented as an alternative to the idea of 'flow' – a term made famous by psychologist Mihaly Csikszentmihalyi, who characterized it as a state of complete immersion in an activity that's based on a subjective feeling rather than a mechanistic, biological process.

One intriguing theory suggests that the direction of visual focus might influence brain states. Proponents propose that directing your eyes upward (for example, towards screens placed at or slightly above eye level) could potentially engage brainstem pathways to heighten alertness. Meanwhile, looking down or reclining might trigger sleepiness through similar mechanisms.

This research reveals a complex reality: until robust evidence confirming this direct brainstem link emerges, posture science continues to suggest optimizing your screen position for physiological reasons. Crucially, established ergonomic guidelines dictate placing the top of your display at or slightly below eye level to maintain a neutral neck position, an approach thought to reduce strain, prevent fatigue and sustain focus. Which approach ultimately wins? Only time will tell.

LEVERAGING YOUR BRAIN'S NATURAL PROCESSES

The brain operates on predictable biological rhythms that smart students can harness. For example, sleep cycles profoundly affect memory consolidation, with the first night after learning being especially critical. Multiple research teams have documented how recent experiences are replayed during sleep, with studies published in *Nature Reviews Neuroscience* and the *Journal of Neuroscience* revealing that this hippocampal replay can persist for several hours following novel experiences. While you're in deep and REM sleep, your brain 'reviews' the day's events. At the same time, bursts of brain activity called sleep spindles act like couriers and construction crews, moving those fresh memories out of temporary storage and weaving them securely into your brain's long-term knowledge base in the cortex.

Circadian rhythms (see page 152) create distinct cognitive phases throughout the day. The first eight hours after waking typically feature higher dopamine and noradrenaline levels, making this period ideal for analytical, precise and detailed work. The following phase, roughly nine to sixteen hours after waking, shifts towards hormonal and neurochemical profiles that favour creative and abstract thinking and brainstorming. So, if you align your study schedules with these natural rhythms, you're effectively working with your biology rather than against it.

Spaced repetition has been shown to be remarkably effective in improving long-term retention and is backed by extensive research that used brain imaging techniques such as electroencephalography (EEG), functional magnetic resonance imaging (fMRI) and behavioural studies. Testing what you have retained soon after exposure to new material significantly increases what you remember compared to passive review (which involves recognizing information rather than actively retrieving it from memory), though the precise magnitude varies across individuals and materials. This technique transcends simple assessment because it actively constructs knowledge. Trying to remember, even if

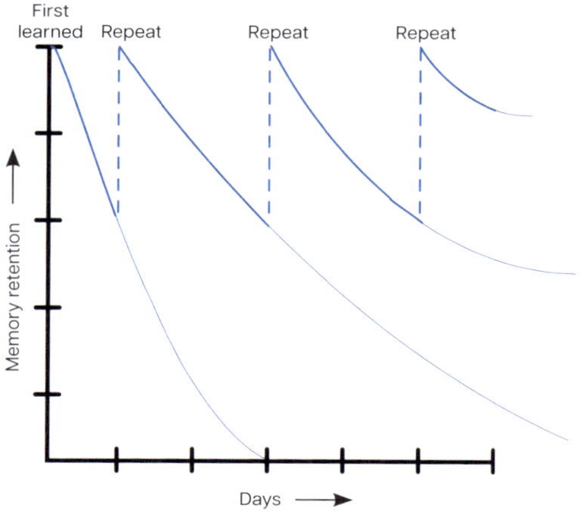

you get it wrong, is neurologically powerful. You essentially train your brain to notice what it needs, strengthen what it has and self-correct over time. That's why flashcards, retrieval practice and spaced repetition are gold-standard techniques in the science of learning.

Some researchers theorize that 'gap effects' represent a powerful biological learning mechanism. According to this theory, brief pauses of five to thirty seconds during active learning cause neurons in the hippocampus to repeat information at rates twenty to thirty times faster than normal, similar to what happens during REM sleep. These proposed micro-breaks create windows for accelerated internal replay and consolidation, acting as spaced repetitions during study sessions. However, the brain's sophisticated consolidation mechanisms are more firmly established through the utilization of strategic rest intervals. While brief breaks during learning do seem to benefit memory formation, the specific timing and neural processes involved continue to be researched. What remains well established is that rest periods and sleep provide crucial windows for memory consolidation, with the hippocampus replaying experiences to strengthen neural pathways and integrate new knowledge.

PRACTICAL IMPLEMENTATION

To really apply these neuroscience findings, you've got to take a close look at your environment and make some thoughtful adjustments.

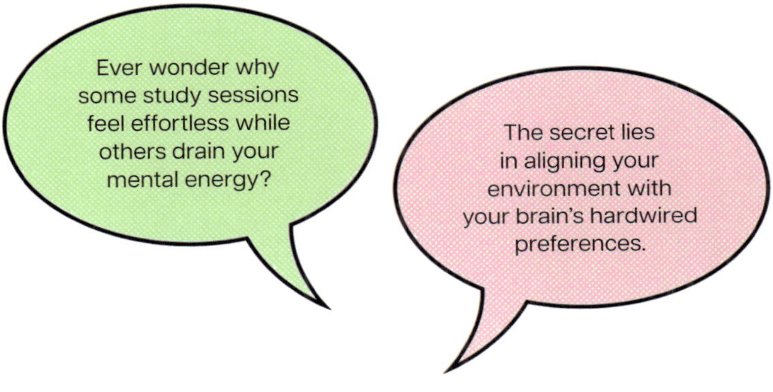

TRY THIS:

ENVIRONMENT AUDIT CHECKLIST

Transform your space by checking off these brain-friendly elements:	
☐ Visual field	Workspace is clutter-free and minimalist
☐ Audio zone	Silence or instrumental music only (lyrics hijack language processing)
☐ Ceiling check	High ceilings for creative work, lower for analytical tasks
☐ Light levels	Bright morning light for analysis, dimmed afternoon light for creativity
☐ Screen position	Adjusted to reduce neck strain and eye fatigue
☐ Rest zones	Designated space for 5–10 minute consolidation breaks

THE PEAK PERFORMANCE TIMELINE

Your brain operates on predictable chemical rhythms – here's when to tackle different tasks:

Time window	Brain chemistry/ state	Optimal tasks	Environmental setup
Morning	Peak dopamine and noradrenaline	Analytical work, problem-solving	Lower ceilings, bright lighting
Midday	Sustained attention	Focused learning, memorization	Consistent environment (controlled and stable workspace)
Afternoon	Declining alertness	Creative work, synthesis	Higher ceilings, softer lighting
Evening	Recovery mode	Review, light practice	Calm, minimal stimulation

THE CONSOLIDATION CYCLE

Here's how your brain processes new information during study breaks:

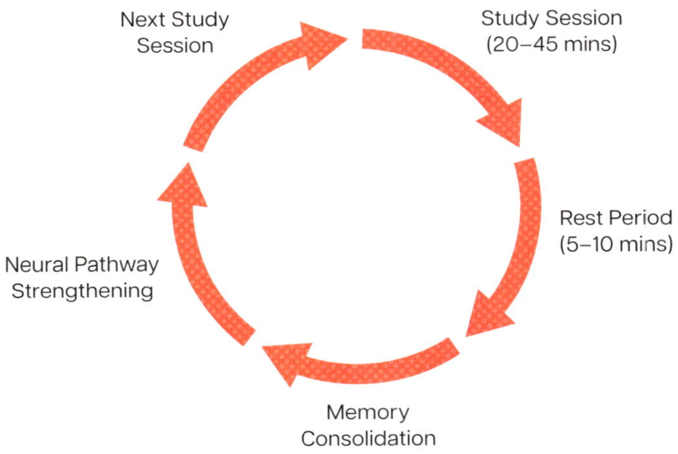

Key insight: Those 'unproductive' rest moments are actually when your brain solidifies new learning.

But how long can you really focus during a study session?

- Ultradian cycles (~90 minute): your brain's natural alertness rhythm
- Focus sweet spot (20–45 minutes): neural fatigue starts to creep in after this period
- Quick reset: Pomodoro Technique[6] ➡ 25-minute work plus 5-minute break
- Deep dive: 90-minute block plus 15–20 minute longer rest
- Tune to you: log your attention dips and errors, then adjust work/break lengths

[6] A simple but powerful time management method designed to boost focus and reduce burnout by breaking work into short, focused intervals.

Become your own neuroscience researcher with this systematic approach

Timeline	Testing phase	Protocol
Week 1–2	Baseline measurement	Track the following during current study habits: • Focus levels – play quick reaction or memory games before and after studying, and rate your concentration on a simple scale immediately after each session • Retention – test yourself immediately after studying; flashcards or simple quizzes work well for tracking this • Energy levels – rate how energetic you feel at different times of the day, or check wearables data like sleep quality from smartwatches or fitness trackers
Week 3–4	Single-variable testing	• Change only your audio environment (silence versus instrumental) • Measure the same metrics as you did during the baseline measurement
Week 5–6	Next variable	• Adjust lighting or ceiling height while keeping audio constant • Continue tracking
Week 7+	Optimization	• Combine your best-performing variables into a personalized protocol

Which single environmental change could you test this week to boost your study effectiveness?

What time of day do you naturally feel most alert for different types of thinking?

QUESTIONS TO EXPLORE

How might you design micro-experiments to discover your brain's unique preferences?

IS MY LIFE BORING?

That's quite a subjective question, isn't it? What you might find tedious and boring about your life another person might find interesting or exciting. Not that you should feel any sort of guilt about feeling bored with the way things are going. It is what it is, and context is everything. So, when you find yourself waking up every morning only to discover it's the same day on repeat – the same commute, the same conversations, the same lukewarm latte and soul-destroying deadlines – day after day, erasing any spark of surprise...well, just like the movie of the same name, you're living a real-life 'Groundhog Day'.

The good news is that you're not alone in wondering whether your life has become mind-numbingly routine – almost everyone will feel this way at some point in their life. And that creeping sense of boredom isn't a character flaw or a sign that your life lacks meaning. Your brain is literally wired to crave surprise, and understanding why reveals both the source of the problem and the solution.

SOMETHING'S NOT RIGHT

Far from being a simple complaint about one's dull circumstances, boredom actually represents a sophisticated neurobiological alarm system. Dr Erin Westgate and Dr Timothy Wilson have developed what they call the MAC (Meaning and Attentional Components) model, which reveals that boredom is a signal that something important is missing: either your attention isn't properly engaged, or your current activity lacks any sense of personal meaning.

When boredom strikes, your brain surprisingly becomes remarkably busy. The default mode network, which is responsible for mind-wandering and self-referential thought, starts firing alongside executive control networks that normally handle focused attention. Meanwhile, reward-related circuits in the striatum, prefrontal cortex (PFC) and anterior cingulate cortex (ACC) become hypersensitive to potential rewards. In practical terms, it's as if your brain is scanning the environment, desperately seeking something – anything – more engaging. You become more driven by (even fixated on) anything that looks like it will offer a payoff. Those pathways that normally register 'good stuff' (the striatum, PFC and ACC) crank up their responsiveness so that even small hints of reward trigger a big surge of neural activity.

This neural restlessness served our ancestors well. From an evolutionary perspective, boredom prevented them from wasting precious mental energy on repetitive, non-informative experiences. Instead, it motivated exploration and learning – crucial behaviours for finding new food sources, avoiding threats and adapting to changing environments. As neuroscientist Dr Charlotte Burn notes from her research on animal behaviour, even laboratory animals exhibit boredom-like states when understimulated, suggesting this drive for novelty runs deep in our biology.

SURPRISE

The key to understanding why routine can feel so demoralizing to some people lies in how your brain processes rewards. Dopamine neurons in the midbrain don't simply signal pleasure; they're sophisticated prediction machines that generate what researchers call 'reward prediction errors'. This means that when outcomes differ from what you were expecting, these neurons fire intensely, creating that ever-so-satisfying surge we associate with pleasant surprises.

Professor Wolfram Schultz, whose landmark research on dopamine reward systems earned him widespread recognition, discovered that unpredictable rewards trigger stronger dopaminergic responses than predictable ones. This explains why a surprise coffee from a colleague feels more pleasurable than your routine morning brew, even if both taste identical. Your brain doesn't just want rewards; it wants to be surprised by them.

The gambling industry has exploited this biological quirk for decades through intermittent reinforcement schedules. Slot machines deliver unpredictable payouts precisely because uncertainty amplifies the brain's reward response, creating that irresistible 'maybe this time' feeling. But the same mechanism that makes gambling so addictive also explains why unpredictable experiences feel inherently more satisfying than routine ones.

Research reveals that this prediction error system involves not just dopamine, but also serotonin and noradrenaline, creating a complex neurochemical cocktail that responds to novelty and surprise. Together, brain areas such as the striatum, PFC, ACC and insula work together to drive our craving for the unexpected, constantly pushing us towards new experiences. It's worth noting that this whole system isn't designed to make you a thrill-seeker. Its main job is to reduce surprise in the long run. By paying close attention to what you didn't expect, your brain builds a more accurate model of the world. This 'craving for the unexpected' is really a drive to explore and learn just enough to make better predictions tomorrow.

EVERYONE'S DIFFERENT

Not everyone experiences boredom equally. Genetic variants in dopamine-related genes, particularly DRD4, influence how intensely people seek novelty. Personality traits like extraversion and openness to experience also predict novelty-seeking behaviour, while brain connectivity patterns in frontal-subcortical circuits shape individual temperament.

Age plays a crucial role too. Adolescents show heightened reward sensitivity and novelty-seeking due to the ongoing maturation of their reward circuits, which explains why teenagers often seem addicted to stimulation. Adults have developed greater executive control, leading to more balanced risk-taking, while older adults sometimes experience declining curiosity. However, research by Dr Michiko Sakaki suggests that maintaining curiosity throughout life supports both cognitive health and mental well-being.

The distinction between healthy and problematic novelty-seeking matters significantly. Beneficial novelty-seeking drives learning, creativity and adaptation. But when it overrides executive control or becomes an end in itself, it can lead to risky behaviours, addiction and other maladaptive patterns.

Healthy novelty-seeking is your brain's way of inviting you to learn, grow and adapt. It sparks curiosity, whether you're eager to master a new instrument, taste an unfamiliar cuisine or tackle an ambitious project at work. In these cases, the prefrontal cortex guides your explorations with planning and self-control, ensuring that excitement fuels progress rather than recklessness. Over time, these positive detours build skills, resilience and confidence.

Your reward circuitry (the striatum and midbrain dopamine cells) registers surprises as 'good' and motivates you to repeat them. Meanwhile, the PFC and ACC normally weigh risks and inhibit impulses. In healthy novelty-seekers, these networks collaborate, but in problematic novelty-seekers, frontal-lobe oversight falters, allowing dopamine's pull to dominate decision-making.

Problematic novelty-seeking emerges when that spark overrides executive control or becomes the primary goal. Without the brakes of your frontal lobes, you may:

- Chase ever-bigger thrills, like extreme sports or high-stakes gambling, despite mounting dangers
- Succumb to substance use or compulsive behaviours because 'just one more' never satisfies
- Sacrifice obligations, relationships or finances in pursuit of the next hit of novelty

Left unchecked, this spiral can lead to addiction, legal troubles or even life-threatening situations.

DO THIS FOR A LESS BORING LIFE

Understanding your brain's hunger for unpredictability opens up practical strategies for combating boredom without upending your entire life. Small changes can create significant neural impact, because your reward system responds to relative novelty, not absolute adventure.

Consider 'micro-adventures' by taking different (safe) routes to work or trying random menu items. These minimal disruptions to routine provide enough unpredictability to engage your prediction error system without causing stress. Research on 'leisure crafting' shows that engaging in varied hobbies and activities outside work can significantly reduce boredom and enhance well-being.

Gamification offers another powerful approach. Adding game-like elements or challenges to routine tasks – whether through apps, personal competitions or skill-building goals – introduces the unpredictability your brain craves. The key is creating situations where outcomes remain uncertain but personally meaningful.

Social novelty counts too. Meeting new people, varying social contexts or even having deeper conversations with familiar people can provide the stimulation your reward circuits seek. The goal isn't constant stimulation, but rather strategic doses of unpredictability within a stable framework.

Mindfulness presents an intriguing alternative approach. Rather than seeking external novelty, mindfulness practices can help you find freshness in familiar experiences by enhancing present-moment awareness. This reduces dependence on external stimulation while still satisfying your brain's need for engagement.

CAUTION NOW

However, the story isn't entirely straightforward. Recent research reveals important complexities that challenge simple dopamine-centric explanations. Primate and rodent studies show that novelty-seeking can engage brain circuits separate from classic reward pathways. The zona incerta and temporal cortex drive novelty-seeking even when novel experiences aren't rewarding, suggesting that curiosity and reward are partially separable neural processes.

Doctors Kent Berridge and Terry Robinson note that 'reward' itself isn't a single unified process but comprises three parallel components: learning, liking and wanting, each governed by distinct neural circuits. Learning involves forming predictions and associations about outcomes, mediated by cortical and hippocampal networks. Liking is the actual feeling of pleasure or enjoyment we get from something, and it's driven by specific areas in the brain's emotional centres that respond to natural chemicals like opioids and cannabinoids. Wanting is the strong urge or motivation to pursue a reward, and it's powered mainly by dopamine circuits that link the midbrain to emotional parts of the brain.

Crucially, dopamine fuels the drive to seek a reward without directly producing pleasure, which arises from other neurochemical systems, and learning depends on yet other circuits. This separation means that dopamine surges can heighten craving ('wanting') without necessarily increasing enjoyment ('liking') or altering learned expectations, challenging simplistic models that tie dopamine to all facets of novelty, pleasure and habit formation.

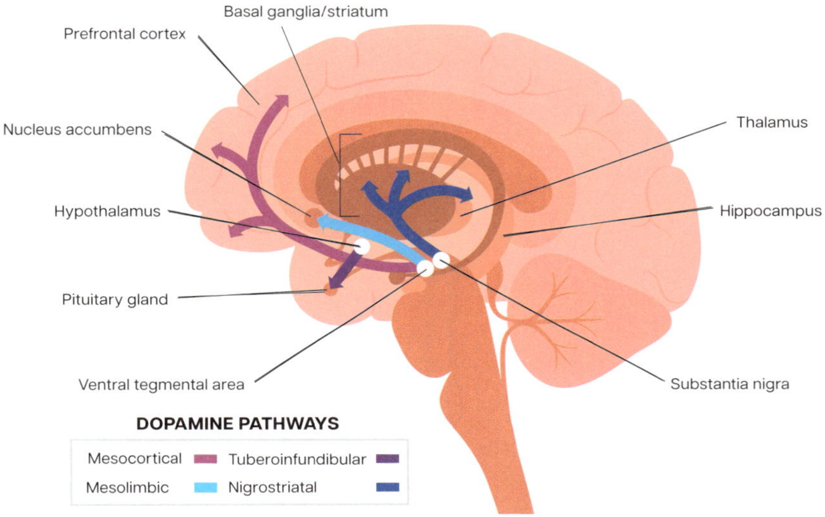

Additionally, boredom research faces significant methodological challenges because boredom is both temporally unstable and deeply context-dependent, making it hard to measure and replicate across studies. As a result, current models often risk oversimplifying what is in fact a complex, dynamic psychological state.

RETHINKING BOREDOM

Boredom isn't proof of a dull life but a signal your brain needs more engagement or meaning or both. Instead of erasing routine, which provides efficiency, security and cognitive resources for when you need them, aim for what researchers call 'optimal balance': maintaining beneficial routines while introducing manageable doses of novelty and unpredictability.

The urge for surprise is an evolutionary drive for growth. Next time you catch yourself thinking, 'My life's so boring,' remember that your brain is simply doing its job by scanning for opportunities to help you learn, grow and engage more fully with the world around you.

ARE YOU ADDICTED TO SUCCESS?

So, you've just landed yourself a major promotion. The rush is intoxicating – your heart races, your mind buzzes with pride. But within days, maybe even hours, that high seems to fade. You're already looking for the next win, the next achievement that will recreate that euphoric feeling. Sound familiar? If so, you might be experiencing 'success addiction' – a term used to describe how your brain's reward system can become so fixated on achievement that it mirrors the patterns seen in drug addiction.

A SUCCESS ADDICT'S BRAIN

Deep inside your brain lies an ancient circuit that evolved millions of years ago to keep our ancestors alive. This system, centred around (but not exclusive to) three key regions, is what drives your pursuit of success today.

First, there's the ventral tegmental area, or VTA – think of it as your brain's motivation engine. When you're chasing a goal, this region pumps out dopamine, a chemical messenger that creates feelings of anticipation and drive. This dopamine floods into the nucleus accumbens (NA), your brain's pleasure centre, creating the rewarding sensation you feel when success is within reach. The third player is your prefrontal cortex (PFC), the brain's CEO. This region helps you plan your next move and decide which achievements are worth pursuing. Together, these three areas, which are part of your reward system's 'mesocorticolimbic dopamine pathway', in a sense form your core 'success circuit'.

Recent studies show that this exact same circuit lights up when people use drugs, gamble compulsively or engage in other addictive behaviours. The overlap is so striking that neuroscientists now draw direct parallels between chasing success and chasing a chemical high. But not everyone agrees with this comparison. Some researchers argue that motivation is far more complex than just dopamine-driven reward-seeking, and that it involves networks related to novelty and genuine curiosity.

HOW THE BRAIN ADAPTS TO ACHIEVEMENT-SEEKING

Here's where success addiction gets its hooks in you. Each time you achieve something significant, your brain adapts. Just like with drugs, repeated exposure to success-induced dopamine leads to tolerance. Your dopamine receptors become less sensitive, meaning you need bigger and bigger wins to feel the same rush. This desire for bigger wins isn't just a psychological thing, though, as your brain actually physically rewires itself. Constantly seeking achievement over long periods of time strengthens the brain's goal-driven pathways (boosting persistence and focus), but it might also sideline circuits linked to savouring experiences. While chronic achievement-seeking doesn't clearly damage self-control circuits, it can create a pattern where you prioritize striving over enjoyment. For many, this leads to a cycle of chasing bigger successes for diminishing satisfaction.

Think of it like a drug user who needs increasingly larger doses to get high. Except, instead of drugs, you're chasing promotions, recognition or financial milestones. The pattern is remarkably similar: escalating effort for diminishing returns. However, it's worth pointing out that not everyone who achieves a lot develops these addiction-like patterns. The human brain's motivation system is more nuanced than a simple reward-seeking machine.

STRESS'S DUAL ROLE

Stress plays a complicated role in success addiction. In small doses, stress hormones such as cortisol can actually boost your performance. They sharpen your focus, enhance your memory and give you the energy to tackle challenging goals.

But when it turns into an obsession, your body never really 'turns off' its stress mode. Instead of a short burst of fight or flight, you end up with a constant flood of cortisol. That hormone overload then begins to rewire your hippocampus, the part of your brain that helps you form and store memories, by shrinking connections, slowing the birth of new brain cells and messing with how cells talk to each other. Meanwhile, your prefrontal cortex, which is in charge of helping you make decisions and curb impulses, starts to run on empty, making it harder to manage your emotions. Ironically, all this brain scrambling can make you even more tuned in to the promise of your next win, so you crave rewards even more.

Pretty soon you're stuck in a loop: the more pressure you put on yourself, the more your stress system fires up, which only breaks down your memory and self-control centres further. To quieten the anxiety or feel like you've got things under control, you push yourself to achieve even more, but this just cranks your stress higher, so you need bigger and bigger successes to get the same hit of relief.

Animal experiments have shown exactly how chronic stress reshapes neural circuits, but in humans we mostly see strong links rather than clear-cut proof of cause and effect, simply because it's not easy to track living brains over long stretches of stress.

IS IT IN THE GENES?

Your genes play a significant role in determining whether you're vulnerable to success addiction, because specific genetic variations affect how sensitive your brain is to dopamine and how well you handle stress: for instance, via variations in dopamine-related genes such as DRD2, DRD4, DAT1 and COMT. These genetic differences are associated with individual variability in reward-seeking behaviours, motivation and stress resilience, and they might contribute to a predisposition towards compulsive achievement or 'success addiction' in some people.

However, genetics is only part of the story. Environmental factors, such as family expectations, workplace culture and societal values, interact with genetic predispositions and can significantly shape behaviour. Through these 'epigenetic' mechanisms, experiences and environments can alter how genes are expressed, further influencing reward sensitivity and stress responses over time. Thus, while certain genetic profiles might increase vulnerability to reward-driven behaviours, environmental and behavioural factors play equally important roles, and genetic predisposition does not determine your destiny.

BRAIN SCANS

If you could peek inside the brain of someone with compulsive reward-seeking behaviours (which success-seeking addicts have) using advanced neuroimaging technology, you'd see a tell-tale pattern. Brain scans show specific changes and disruptions in reward-related regions such as the striatum, orbitofrontal cortex and medial prefrontal cortex whenever these individuals anticipate or achieve a goal. The pattern is remarkably similar to what researchers see in people with substance addictions or gambling problems.

Some studies suggest that chronic engagement in reward-chasing might actually change the physical structure of your brain, altering grey matter volume in regions responsible for impulse control and emotional regulation. However, these findings are more robust in substance-use disorders than in behavioural addictions, and whether these changes are a cause or a consequence of compulsive behaviours is still being debated. Larger, longer-term studies are needed to confirm these structural changes.

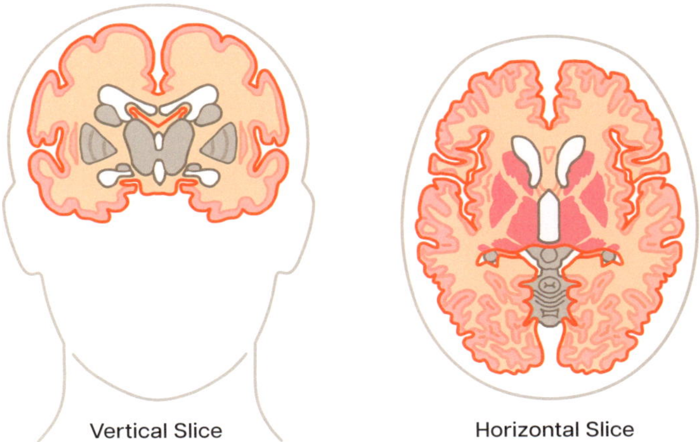

Vertical and horizontal brain slices showing the outer cortex and inner brain structures commonly examined in behavioural addiction brain scans.[7]

---

[7] Schematic illustration – not an actual scan.

TRY THIS:

YOUR BODY KEEPS SCORE

Success addiction isn't just a head thing, because it leaves measurable traces throughout your body. And studies show that chronic stress from achievement pressure can show up in various ways.

'The hidden cost of high achievement' self-assessment

(0 = Never | 1 = Sometimes | 2 = Often | 3 = Always)

Answer honestly: How often in the past three months did you...

YOUR DRIVE

1. Cancel rest or social plans to work on goals?
   (Relevance: Cortisol disrupts prefrontal cortex ➡ impaired work–life balance)
2. Feel empty after achieving a major milestone?
   (Dopamine receptor downregulation ➡ diminished reward sensitivity)
3. Need bigger 'wins' to feel satisfied versus one year ago?
   (Neuroadaptation in nucleus accumbens ➡ tolerance development)

YOUR BODY

4. Wake exhausted despite 'enough' sleep?
   (Chronic cortisol ➡ disrupted REM/deep sleep cycles)
5. Notice unexplained tiredness/muscle joint aches/bloating nausea issues?
   (Stress-induced inflammation ➡ elevated levels of inflammatory markers IL-6/CRP)
6. Feel physically 'wired but tired' (heart racing at rest)?
   (Low heart rate variability ➡ autonomic nervous system dysregulation)

YOUR EMOTIONS

7. Snap at colleagues/family over minor issues?
   (Prefrontal cortex (PFC) impairment ➡ reduced emotional control)
8. Feel guilty when not being productive?
   (Amygdala hyperactivity ➡ anxiety-driven behaviour reinforcement)
9. Believe 'I'll relax when...' (but never do)?
   (Dorsal striatum habit loops ➡ compulsive goal pursuit)

Scoring

Add all points ➡ See your risk level:

KEY PHYSIOLOGY BEHIND YOUR SCORE	
Score range	**What's likely happening in your body**
Low (0–9)	Cortisol rhythms intact ➡ Healthy heart rate variability (HRV) ➡ Balanced dopamine/serotonin
Moderate (10–18)	Cortisol spikes persisting >1 hour ➡ Reduced HRV ➡ Early dopamine receptor changes
High (19–27)	Flat cortisol curve (AM/PM) ➡ Chronically low HRV ➡ Reward circuit tolerance plus PFC impairment

This very casual tool screens for potential physiological patterns linked to chronic-achievement stress. For a definitive clinical diagnosis, consult a healthcare provider.

Scientists are working to develop reliable biological tests for addiction behaviours, but validated measures are still in the early stages.

BREAKING FREE

The encouraging news is that your brain's ability to change (neuroplasticity) works in both directions. Many of the addiction-like changes associated with compulsive achievement-seeking can be reversed with the right approach.

Behavioural strategies (foundation of recovery)	Mindfulness practices	Regulate runaway stress responses
	Cognitive techniques	Challenge the inner voice that equates your worth with your achievements
	Scheduled downtime	Allow your dopamine system to reset
	Cultivating rewarding activities	Provide alternative sources of fulfilment outside work, e.g. hobbies, relationships and creative pursuits
Medicines	Antidepressants	Treat anxiety and depression
	Addiction medications (e.g. naltrexone)	Reduce reward cravings

Being addicted to success represents a fascinating consequence of our ancient brain circuits meeting modern-day pressures and expectations. While the drive to achieve has served humanity well throughout evolution, our current culture of constant competition and endless growth may be pushing these systems beyond their healthy limits.

The key, as always, is finding balance by maintaining the motivation to grow and accomplish meaningful things while preserving your mental health, relationships and overall well-being. Remember, your brain's reward system evolved to help you thrive, not just survive the endless pursuit of the next 'big win'.

THE STRESS ADAPTATION PARADOX

Everyone gets stressed at some point in their life – it's part of being human. Stress is a bit like having a weather system inside you. Sometimes it's a brief summer storm – loud, maybe a bit thrilling, but over quickly. Other times it's a grey, unending drizzle that soaks into and slowly destroys everything.

Biologists and psychologists talk about stress as being a state in which your body's balance, or homeostasis, is threatened. Now, that threat can be either real, like a speeding car swerving towards you, or something you perceive only in your mind, like the thought of an awkward conversation with your boss, for instance. Different fields within science debate whether to define stress by the thing causing it (the stressor), or by your inner reaction, or even by your subjective feeling about it.

Two main varieties of stress get most of the attention: acute and chronic. Acute stress is the quick flash. Chronic stress is the slow burn. The first can be useful in the right circumstances; the second, damaging.

QUICK EQUALS GOOD

Acute stress is your body's immediate rally to a challenge. Think of things like cold showers, high-intensity workouts, intermittent fasting or resistance training. These are examples of hormetic stressors: brief, controlled challenges that actually strengthen the body over time by triggering protective adaptations.

When a threat appears, your senses or thoughts swiftly detect it. Your hypothalamus kicks into gear, which triggers the hypothalamic–pituitary–adrenal (HPA) axis and the sympathetic nervous system. This floods your system with hormones such as cortisol, and catecholamines like adrenaline and noradrenaline released from the adrenal medulla, which drive broad stress responses, while noradrenaline released from nerve endings of the sympathetic nervous system allow for quick and targeted control of body functions.

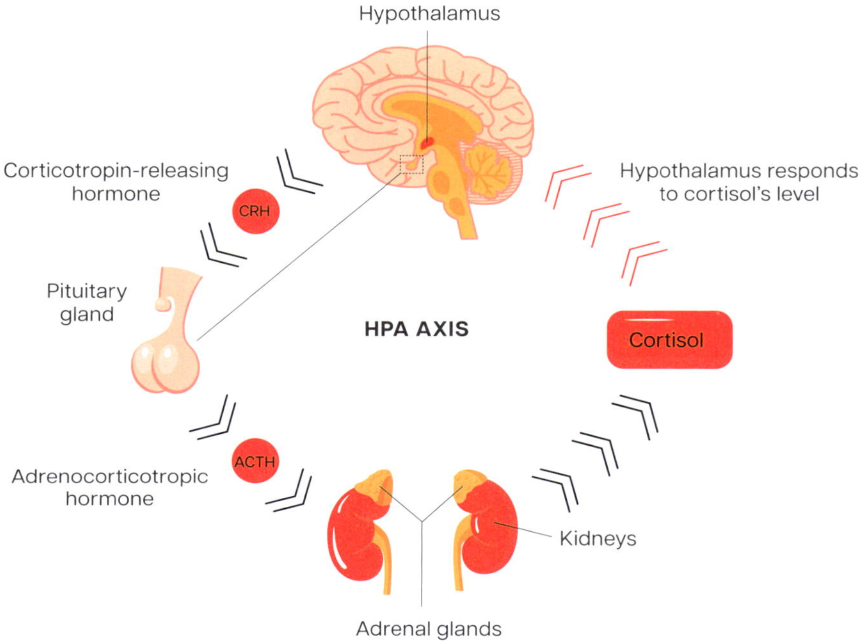

These chemical messengers mobilize the body's resources, prepping for fight or flight. The heart pumps harder. Muscles get more blood. Your mind sharpens. Even your cells shift their protein-making priorities to handle the situation. Psychologically, it often feels like a sharp, solvable problem. A deadline looms, you focus and power through, and – if it ends there – you might come out sharper for it.

In short bursts, this kind of stress can lift your cognitive game. Studies have shown it can improve memory recall, decision-making speed and attention span. Athletes sometimes perform their best under these circumstances, referred to as so-called 'clutch' moments. It can also train resilience, thanks to the brain's ability to rewire itself. But there's a catch: the dose matters. Too strong or too long, and the benefits melt away and easily tip into harm.

SLOW EQUALS BAD

Chronic stress is the one to watch for. This is stress that doesn't quit. Your system stays in high gear for weeks, months or even years. The HPA axis remains activated, so the stress hormones persist in your bloodstream.

Cortisol levels remain high

Natural rhythms are disrupted

Brain areas such as the hippocampus, amygdala and prefrontal cortex start to change, and they sometimes shrink or over-fire

Immune system misbehaves

Inflammation lingers

Risk increases for a whole range of illnesses: heart disease, diabetes, depression and even dementia

Chronic stress also has an impact at the cellular level:

- Mitochondrial dysfunction: cells lose efficient energy production
- Impaired protein folding: molecular machinery stalls
- Disrupted DNA repair: mutations accumulate

Scientists call the accumulated wear and tear from chronic stress allostatic load. It's as if the weather system inside you never clears, eroding the landscape bit by bit. Mechanisms that once helped, such as quick energy release, become corrosive over time, damaging cells, fraying repair processes, and leaving you more vulnerable to infections and disease.

Now, your body does have an 'off' switch for stress. It works through feedback: cortisol binds to receptors in your brain, signalling them to stop the alarm. Other brain circuits help quieten the stress response. But with chronic stress, this machinery breaks down. Feedback loops lose sensitivity, and brain regions that regulate stress are damaged. It's like a thermostat that can no longer sense the room temperature, so the heating runs endlessly.

This paradox arises because the pathways of acute and chronic stress are nearly identical. In the short term, activation triggers adaptive responses. Prolonged activation, however, overwhelms those same responses, turning protection into pathology. The tipping point depends on how long, how intensely and how uncontrollably stress hits.

INDIVIDUAL DIFFERENCES

Of course, not everyone reacts to stress in the same way. Your genetics matter – some people inherit more sensitive stress circuits. Your early life matters, too – childhood adversity, for example, can prime the stress response to be more reactive later in life. Personality, coping style and social support all play their part. Researchers have even linked these differences to epigenetic changes – chemical tweaks to DNA expression that alter how your cells respond.

PANORAMIC VISIONING

The good news is you can train your system to be more adaptable. Exercise, proper sleep and balanced food support healthy stress responses. Psychological tools such as mindfulness, therapy or simply having strong social connections help too. For more severe cases, medication may be the only way to help you first get off the spinning wheel of doom.

One intriguing stress-relieving method from neuroscience comes in the form of panoramic vision. Dr Andrew Huberman, a neuroscientist and professor at Stanford University, is among the first to popularize the scientifically established health benefits of panoramic visioning through his work on the visual system's link to autonomic arousal and stress regulation.

PANORAMIC VISIONING

A simple exercise to calm your nervous system and reset your stress response.

- Step outside and find a wide view, ideally a horizon, open field or long street.
- Soften your gaze so you're taking in the whole scene, including your peripheral vision. You're not locking on to a single point; instead, let your view widen without moving your head.
- Let your eyes wander naturally across the space, sensing the edges of your visual field – to the sides, and above and below.
- Add movement for extra benefit; walking while doing this engages 'optic flow', a visual pattern that further soothes the nervous system.

WHY IT WORKS:

- Activates neural pathways that cue a shift from high alert sympathetic mode into the calmer parasympathetic state.
- Reduces cortisol, relaxes your eye muscles and calms the mind.
- Particularly effective after long periods of screen use.
- **TIP:** even a minute or two can make a noticeable difference, though longer sessions deepen the effect.

A panorama that would be ideal to help shift you from tunnel to peripheral vision and induce calm.

Dr Huberman has described panoramic vision as a tool to tune your nervous system. Focused vision is like an 'on' switch for drive and productivity, whereas panoramic vision is the 'off' switch – not to sleep, but to restore calm and reset your attention. While large-scale clinical trials on panoramic visioning are still emerging, preliminary neuroscience findings suggest it might support autonomic balance by activating the parasympathetic nervous system. The proposed mechanism (engagement of the superior colliculus to dampen stress pathways) is mechanistically plausible and aligns with evolutionary biology. However, as with many neuromodulation techniques, individual responses can vary, and ongoing research aims to further confirm that these protocols actually produce reliable, repeatable results. For now, its minimal risk, ease of practice and grounding effects make it a compelling adjunct to existing stress-management tools.

THINGS TO DIGEST

This sounds promising, but as with much of science, the reality is a bit more nuanced. Some people's stress systems adapt quickly to repeated exposure, while others don't. For example, cortisol responses might dampen with repetition, but immune reactions can stay high or even get more intense. This means some people continue to carry a heavy allostatic load even after they've 'become used to' the stressor.

There's also the concept of hormesis discussed earlier (see page 145). A little bit of stress can be good, because it trains your systems to adapt, but too much tips into harm. The optimal dose depends on who you are, what the stressor is and your current state. It's not a one-size-fits-all situation.

And then some evolutionary models suggest that even prolonged stress can be adaptive in the right environment. For example, early stress, which refers to significant and persistent adversity during childhood, such as neglect, abuse or chronic family conflict, is often considered chronic because, for a developing child, such adverse conditions are typically not one-time events but a sustained reality. And because it occurs during a critical window of development it doesn't just cause a temporary reaction; it can programme the system to be permanently on high alert. This persistent state programmes the body's stress-response systems to adapt for survival, leading to biological strategies like accelerated puberty to enhance reproductive chances in a perceived high-risk environment. Such strategies could help in dangerous settings, even if they come with long-term health costs. Survival, after all, doesn't always mean thriving.

Misunderstandings about stress are common. One is that all stress is bad. Not so! Without acute stress, you'd lose an important survival tool. Another is that stress is purely in the mind. In reality, it's an intricate mix of biology, psychology and environment.

Researchers are still chasing big questions. What exactly tips the balance from acute to chronic stress? Can we pinpoint biomarkers for resilience? How do modern stressors, such as those constant phone notifications, screen glare and urban noise, interact with ancient stress systems? The answers could shape personalized interventions in the future.

Stress is nature's performance booster and slow-acting saboteur. To thrive, you must learn to harness the benefits of acute stress while avoiding the corrosive effects of chronic exposure. The key lies in understanding how your body and mind mediate this delicate balance, and designing lives that let adaptation work for us, not against us.

WHY YOU NEED TO CATCH THE MORNING (AND EVENING) SUN

The interaction between natural light and your body's internal clock is such a fundamental biological process, and yet most people often neglect it and don't think much about it. However, this daily light–dark cycle plays a vital role in helping you sleep well, maintain a balanced mood and support overall health. It's the primary signal that keeps your body's internal rhythms in sync, but these days, our indoor routines and constant artificial light are quietly messing with that natural harmony. By understanding the profound influence of both morning and evening light on your physiology, you'll have a simple yet potent bio-hack for massively improving your well-being.

THE CIRCADIAN RHYTHM

Every cell in your body operates on an internal timekeeper called your circadian rhythm – a roughly 24-hour cycle that governs virtually every physiological process, including when you feel alert, when you get hungry, hormone release, body temperature, immune function and metabolism. This biological operating system is orchestrated by a master control centre in your brain known as the suprachiasmatic nucleus (SCN), which contains about 20,000 neurons functioning as your central pacemaker.

HOW YOUR BODY'S CLOCKS KEEP TIME

Specific genes (CLOCK, BMAL1, PER, CRY, etc.) turn each other 'on' and 'off' in a pattern that takes about 24 hours to complete one cycle. The output of one part of the loop regulates another, creating a self-sustaining oscillation (feedback loop).

MASTER CLOCK

- Located in the brain's SCN
- Directly reset by light signals detected by the eyes

PERIPHERAL CLOCKS

- Exist in every tissue (liver, heart, lungs, skin, etc.)
- Run autonomously via molecular gears (the core clock proteins produced by genes mentioned above)
- Cannot be reset by light directly

SCN'S SYNCHRONIZING SIGNALS

(How the SCN sends timing 'messages' out to the rest of your body so that all your organs keep the same 24-hour schedule)

- Neural pathways (stimulate key brain regions ➡ sympathetic nervous system)
- Hormonal pulses (for example, melatonin, cortisol)
- Behavioural cues (feeding/fasting, body temperature)

Together, the SCN – the central conductor – and its peripheral 'subordinate' clocks align your entire body to the external day–night cycle.

The SCN doesn't run on its own; it requires external cues, called zeitgebers (or 'time-givers'), to stay synchronized with the outside world. Light is the most powerful zeitgeber of all, and when you're exposed to it is enormously important. Morning light tells your body that it's time to be awake and alert, while evening light signals that you should start preparing for rest. If these signals are wrong, your entire biological system can fall out of sync.

A pivotal breakthrough in understanding light's influence came in 2002 when researchers discovered specialized cells in the back of your eyes called intrinsically photosensitive retinal ganglion cells (ipRGCs). These cells contain a photopigment called melanopsin, which was first identified by Professor Ignacio Provencio and colleagues in 1998. Unlike the rods and cones responsible for vision, ipRGCs (aka melanopsin cells) are your body's light detectors for regulating circadian rhythms. This discovery is remarkable because these cells operate independently of conventional vision; even people who are completely blind are thought to still maintain normal sleep–wake cycles through these melanopsin-containing cells, which send signals directly

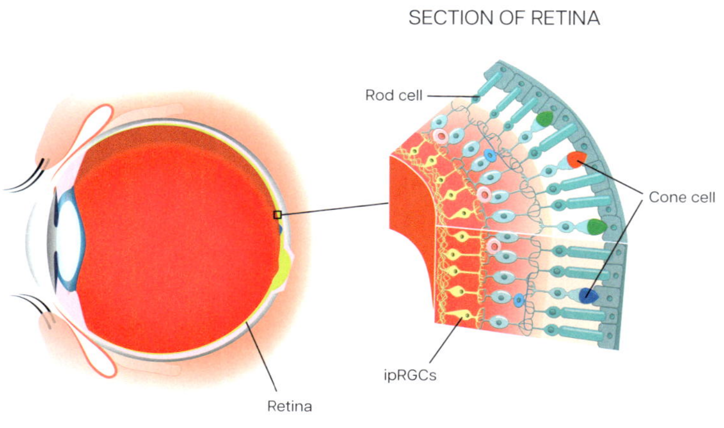

ipRGCs containing the melanopsin pigment located on the innermost layer of the retina.

to your SCN via the retinohypothalamic tract. These cells are designed to detect blue-rich light – a high-energy, short wavelength portion of the visible spectrum ranging from approximately 400–500 nanometres (nm). To put that in perspective, a single metre contains one billion nanometres. They exhibit a peak sensitivity around 480nm.

THE MORNING SUNLIGHT

When morning sunlight hits your eyes (and no, you don't need to look directly at the sun, which should be avoided anyway – just being outside is enough), it triggers a cascade of biological events that prepare your body for active daytime functioning. The low but rising intensity of dawn light combined with your heightened morning light sensitivity means even modest blue-light content can powerfully activate melanopsin cells, signalling your SCN to suppress melatonin and enhance cortisol production. This initiates what researchers call the cortisol awakening response (CAR) – a sharp 38 to 75 per cent increase in cortisol that naturally occurs within 30 to 45 minutes of waking. This is not stress; rather, it's your body's way of mobilizing energy for the day ahead, and boosting alertness, immunity, focus and metabolism.

Surprisingly, ipRGCs can fire in response to light as low as 1 lux, meaning the gentle rise of dawn engages your clock, even though its proportion of blue light is lower than midday's much brighter spectrum. However, because your system is most biologically primed at that hour, a little light goes a long way. That said, some studies have found that waking doesn't always trigger a big cortisol spike, which reminds us that biology isn't one size fits all.

Lux is the standard international (SI) unit of illuminance. It tells you how bright a surface is by counting how many light particles hit each square metre; more lux means the surface looks brighter.

Benefits of morning light exposure include enhanced alertness, improved mood, better cognitive performance throughout the day, and increased energy levels and overall well-being.

Controlled studies show that even half an hour of blue light in the morning can shape your sleep that night, helping you to spend less time in light sleep and reach REM faster.

THE EVENING LIGHT

As the sun sets, physics becomes your ally in promoting a good night's sleep, and that sunset stroll might be what helps you nod off at night. Here's why: during daytime, a phenomenon known as Rayleigh scattering causes the atmosphere to scatter blue light (380–500nm) in all directions. This is the reason why: (i) the sky appears blue; and (ii) you don't need to be looking directly at the sun in the early morning to activate melanopsin cells – scattered blue light from the entire sky suffices.

At sunset, however, sunlight takes a longer path through the atmosphere due to the sun's low angle. This extended journey scatters nearly all of the blue light away from your line of sight, leaving only the longer warm red/orange wavelengths (600–700nm) to reach your eyes directly. Biologically, this matters, because melanopsin cells in your eyes respond weakly to these warm hues. With reduced stimulation, your brain's master clock (SCN) signals your pineal gland to release the sleep hormone melatonin to start preparing you for sleep. Natural evening light with reduced blue content thus protects this critical wind-down rhythm.

Recent research also suggests that natural sunlight contains substantial near-infrared radiation, which some studies propose might actually enhance melatonin production in peripheral tissues. This represents a fundamental difference between natural evening light and artificial sources, which typically lack significant near-infrared content.

Unfortunately, modern life dramatically disrupts the production of melatonin. Typical indoor LED lighting contains substantial blue wavelengths, making it highly effective at suppressing melatonin production, even at moderate intensities. Studies demonstrate that exposure to just 100 to 300 lux of typical indoor lighting during evening hours can delay melatonin onset by 30 to 90 minutes and reduce overall production by 30 to 50 per cent. Even modest increases in evening light exposure above 10 lux correlate with later sleep timing and increased sleep fragmentation.

Light property	Morning sun	Evening sun
Wavelength	Blue (~480nm)	Red/orange (600–700nm)
Melanopsin stimulation	⬆⬆⬆⬆⬆⬆ (Strong)	------ (Minimal)

Sunset's red/orange light fails to activate ipRGCs/ melanopsin cells significantly, allowing:

- Melatonin to rise naturally
- Your body to initiate 'wind-down' physiology
- Sleep pressure to build

WHY THE TIMING OF NATURAL LIGHT EXPOSURE MATTERS

Getting your light cues right is a fundamental health choice. Proper light exposure patterns help keep your body clock strong and steady, leading to more consolidated sleep and improved daytime alertness. On the flip side, getting light at the wrong times (out of sync light cues) is thought to create a chain reaction of health problems. Chronic circadian disruption is increasingly linked to serious health conditions, including metabolic disorders, cardiovascular disease, immune dysfunction, insomnia and accelerated ageing.

OPTIMIZING LIGHT EXPOSURE FOR CIRCADIAN HEALTH

Time of day	Key recommendation	Light level/ type	Actionable strategies
Daytime	Maximize natural light exposure: 15–30 min of direct natural light within the first hour of awakening	≥250 lux	• Catch the sun between dawn and before 8am (sweet spot) • Take lunch breaks outdoors • Sit near open windows • Use bright overhead lighting • Avoid sunglasses or closed windows, as glass filters out crucial circadian-effective blue light
Evening	Begin light reduction 3+ hr before bed	≤10 lux	• Use 2,700K bulbs • Wear blue-light glasses for screens • Embrace natural sunset light • Avoid overhead light • Use low-positioned table/floor lamps
Sleep	Minimize all light exposure	<1 lux ambient light	• Install blackout curtains • Eliminate light sources (LEDs, chargers) • Consider a sleep mask

LUX COMPARISONS:

- Moonlight: 0.1–1 lux
- Home lighting: 100–300 lux
- Office lighting: 500 lux
- Sunny day: 10,000–100,000 lux

THE LUNGS

BREATH, HARMONY AND INTEROCEPTION

5.

AM I BREATHING WRONG?

It's the first thing you do when you're born that reassures everyone that you're alive – that you've made it. From your very first breath as a baby to the last one you draw, breathing is the one thing you do instinctively, without much effort or conscious thought. The problem is, though, between breath number one and your final breath, most of us don't know how to do it well.

TRY THIS:

Take a moment – are you thinking about your breath right now? Most of the time, breathing is one of those quiet backstage things, just happening, no rehearsal required. Yet emerging research says the way you breathe might matter a lot more than you assume.

The simple question, 'Am I breathing wrong?' sounds almost laughable, because how could such an elemental, automatic act be going badly for millions of people every day? But modern science is quietly rewriting what we thought we knew about pulmonary ventilation. There's mounting evidence that there's a 'right' way to breathe, and breathing the wrong way can have a negative impact on your health, your moods and even your ability to think clearly. With busy lives, screens at every turn and rising rates of chronic stress, most of us have never been taught 'good breathing'. A few invisible habits that are barely noticed might be silently shaping how healthy, focused or energized you feel each day.

WHAT IS THE 'RIGHT' WAY TO BREATHE?

When you think about breathing well, most people automatically assume all you need to do is take a big gasp in through the mouth with your chest rising and 'voila' – job done. But that's not exactly a great technique. Decades of research and review papers suggest there is a an optimal way to get air into your lungs, and it has a few basic requirements:

- Nasal
- Diaphragmatic
- Slow

Ideally, and assuming you're healthy, six to ten breaths per minute is what you should be aiming for.

Nasal breathing

Granted, not everyone has the luxury of breathing easily through their nose, with deviated septums and chronic rhinitis two of many ailments that make it hard. But, for those without any issues, nasal breathing confers so many advantages.

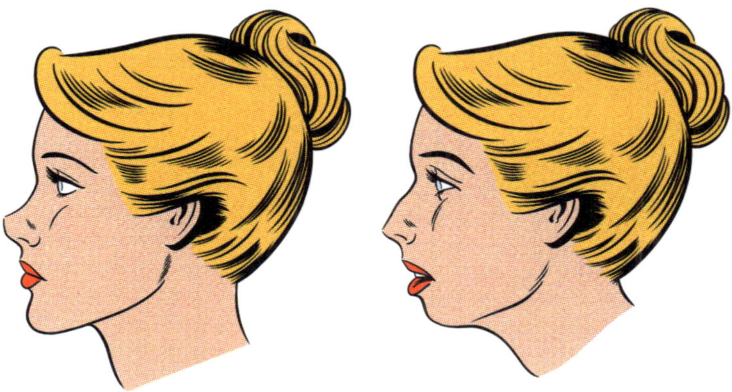

Nasal versus mouth breathing. Chronic mouth breathing can lead to noticeable changes in the structure of your face over time: a narrower upper jaw, receding chin and jaw, and a long, narrow face.

For a start, it transforms air in ways that make every breath healthier and more efficient. When you breathe through your nose, the nasal passages:

- filter out dust and microbes,
- humidify and warm the air,
- help your lungs to absorb oxygen better.

They do the last one by boosting your intake of nitric oxide, a simple gas produced in the nasal cavity that opens up blood vessels and makes oxygen transfer smoother as it enters your bloodstream.

On a mechanical level, breathing through the nose encourages deeper, slower breaths that activate the diaphragm. Scientists have also shown that nasal breathing uniquely signals the brain, connecting to areas that affect memory, attention and emotional regulation. In doing so it acts as a gatekeeper that sets the stage for your body and mind to operate in harmony, helping to cultivate a healthier, calmer and more focused self.

Diaphragmatic breathing

That broad, dome-shaped muscle called your diaphragm, which separates your lungs and heart from your abdominal organs, should take centre stage here. That's because breathing deeply with it means your lungs fill more efficiently through the suction effect that causes the ribs to expand gently outward. This approach not only maximizes how much air your lungs can use but also reduces the strain on those smaller chest muscles that otherwise would work hard to expand your chest.

Slow breathing

The top tier of evidence, mainly from large studies of randomized control trials in the general population, shows that low, deep and deliberate breathing amplifies the calming rest-and-digest arm of your parasympathetic nervous system. People practising this style consistently show increases in heart rate variability, which is a good thing, because it means your heart is more resilient to stress and anxiety.

BENEFITS OF GOOD BREATHING:

- lowered stress
- improved mood
- better sleep
- enhanced athletic performance
- alleviated symptoms of asthma
- reduced anxiety
- assisted recovery from respiratory infections; for example, COVID

BREATHING WRONG

So, what does 'wrong' look like? Most of us, without realizing, breathe mainly through the mouth, take fast and shallow breaths, and neglect to use our diaphragms well (aka thoracic or chest breathing). This isn't unusual, as studies suggest that up to 45 per cent of even well-trained athletes show one or more dysfunctional breathing patterns.

This is often due to your anatomy or the state of your nasal passages. For example, a blocked nose from a cold or allergies can push you toward mouth breathing, as can structural issues like a recessed or narrow jaw, which restrict airflow, alter tongue posture and narrow the palate. But a great chunk of it boils down to behaviour and environment: for instance, stress that triggers short, upper-chest breathing or growing up in places with high pollution.

It's tempting to see dysfunctional breathing as a 'you' problem due to poor habits and bad posture. But the environment shoulders much of the blame. Modern life provides the perfect storm: hours spent sitting down, pollution particles in every city breath, chronic work stress, allergies and digital distractions that quietly shorten each inhale. Increased screen time also changes the way we sit and move, feeding poor patterns. Just as our eating habits are shaped by our surroundings, so too are our breaths. But these patterns aren't set in stone; with awareness and support, habits can change.

TRY THIS:

HOW TO KNOW IF YOU'RE BREATHING 'WRONG': A DIAGNOSIS AND SELF-ASSESSMENT GUIDE

While breathing feels automatic, subtle dysfunction can creep in over months or years. The best tools for recognizing this are simple self-checks and reflective questions. Early recognition helps you seek appropriate breathing retraining or medical evaluation before patterns become deeply ingrained.

1. Common early signs

Watch out for symptoms that develop gradually and 'seem to be just a part of you':

- Breathlessness with mild exertion (for example, climbing a flight of stairs)
- Unexplained fatigue or feeling 'worn out' by routine tasks
- Frequent sighs or yawns, especially when you're not tired
- Muscle tension/pain, especially around the neck, shoulders or chest
- Anxiety or irritability that seems to spike without a clear trigger
- Subpar exercise performance despite training

2. Self-check questions

Answer each honestly to see where you might need to focus:

	YES/NO
1. Do you breathe mostly through your mouth, not your nose?	
2. At rest, is your chest rising before or more than your belly when you inhale? (Note: healthy breathing is context dependent. During exercise/stress, chest movement is normal. At rest, aim for belly rise to start each breath.)	
3. Do you notice tightness or tension around your neck or shoulders after breathing?	
4. Do you find yourself holding your breath or taking sudden 'gasps'?	
5. Are you often short of breath during gentle activities?	
6. Do you feel fatigued even when you haven't exerted yourself?	
7. Do you experience frequent sighs or deep yawns without being sleepy?	
8. Have you had random bouts of anxiety that ease when you take a few slow breaths? (Note: anxiety may temporarily ease with controlled breathing.)	

Interpretation:

- **0–2 'yes' answers:** likely normal breathing patterns but look out for new symptoms developing.
- **3–5 'yes' answers:** mild breathing imbalance – consider simple nasal-breathing drills or gentle diaphragmatic exercises.
- **6–8 'yes' answers:** signs of dysfunctional breathing – seek a professional assessment or structured breathing retraining.

3. Quick at-home exercises

If your self-check suggests mild issues, try these techniques on a daily basis (check with your GP first):

1. Nasal-only breathing: inhale/exhale slowly and fully through your nose for 5 minutes.
2. Diaphragm awareness: place one hand on your belly and one on your chest. Breathe so your belly rises first, followed by slight outward rib expansion.
3. Paced breathing: breathe in for a count of 4 seconds, hold for 2 to 4 seconds and breathe out for a count of 6 seconds. Repeat ten times. Stimulates your PNS.
4. Posture check: slouching compresses the diaphragm, reducing lung expansion. Sit tall with your shoulders relaxed and chest open – breathe and note any tension.

4. When to seek professional help

See a specialist if you experience:

- Persistent or worsening symptoms despite home exercises
- Chest pain, dizziness or palpitations during breathing checks[8]
- High levels of anxiety tied closely to breathlessness
- Unexplained fatigue interfering with daily life

There's no single 'gold standard' home test yet, but careful self-observation and simple drills can flag early issues. By tuning in to how you breathe – nose versus mouth, chest versus belly, easily versus tensely – you can feel more confident to correct dysfunctional patterns before they become chronic.

[8] If you experience chest pain or dizziness during any breathing exercises or unexplained fatigue, immediate medical evaluation is advised.

REASONS WHY YOU SHOULD ESCAPE TO THE COUNTRYSIDE

Are you a townie or country person? If the latter, you'll find this reassuring. But if you're a staunch city-lover, go and grab yourself a drink, switch your phone on silent mode and get ready to absorb some reasons why you should get out more to the countryside and immerse yourself in some greenery.

Correction: it doesn't have to be the countryside at all. A big expanse of green with trees aplenty, like a city park, for example, will absolutely suffice if a more rural landscape just isn't your thing.

When you get there, step under a leafy canopy, take a deep breath and let that freshly made air completely fill your lungs. Notice how the world tilts gently, inviting you to pause, unwind and reconnect with your physical surroundings.

Spending time in nature is a simple, affordable way to enhance your physical, mental and cognitive well-being. The benefits are measurable: stress hormones decrease, attention becomes steadier and your immune system might receive a temporary boost too. Science is piecing together how this happens, from the gentle near-infrared light reflected by leaves to aromatic plant compounds called phytoncides all working together to support your health. None of these are magic bullets, and each comes with caveats, but together they help explain why a countryside walk can sometimes feel like a biological reset.

MULTISENSORY RESTORATION

The countryside acts on our senses first, which is why you need to engage them all to reap maximum benefit. The sights, sounds, smells and different textures of nature's plant ware provide the perfect sensory recalibration experience.

Evidence and examples	Benefit	How it works
Birdsong, rustling leaves or water sounds (especially when you can see the source) help you to feel deeply restored	Calming	Reduces stress and recharges you
Spending time in nature reliably lowers cortisol in saliva, showing your body's stress systems are easing	Lowers stress hormones	Physically calms your body down
• Adults in parks often feel clearer focus and solve problems better • Kids with regular access to green spaces tend to have better thinking skills and fewer ADHD symptoms	Sharper thinking	Improves focus and brain development

ALL ABOUT THE GREEN

Healthy plants and leaves strongly reflect near-infrared light (NIR), an invisible part of sunlight, due to their internal structure and chlorophyll arrangement. This reflectance helps plants stay cooler and avoid heat damage, and it is so distinctive that satellites use NIR signatures to assess forest health.

Under a tree canopy, the ambient light field is enriched with scattered NIR mixed with shade. This NIR is far weaker and more variable than direct sunlight, with its intensity shaped by weather, time of day, canopy density and other factors. The result is a gentle, skin-surface exposure to NIR for anyone in a green space, particularly in the early morning and late afternoon.

When specific red and near-infrared wavelengths (typically 660 to 850 nanometres) reach human skin, they can penetrate roughly 0.5 to 10 millimetres into tissue. Pioneering work by Estonian biophysicist Dr Tiina Karu and colleagues established that mitochondria absorb these wavelengths via an enzyme called cytochrome c oxidase. This interaction can temporarily boost cellular energy (ATP) production and increase the synthesis of local antioxidants, including extrapineal melatonin: the version generated within cells throughout the body (skin, gut, immune cells) rather than by the pineal gland. In tissues, extrapineal melatonin acts as a potent antioxidant and anti-inflammatory agent.

This light-driven process, known as photobiomodulation, has been shown in controlled laboratory and clinical settings (using precisely dosed, high-intensity medical devices) to:

- accelerate wound healing,
- reduce pain,
- decrease inflammation.

Modern wound-healing research confirms that when parameters are carefully tuned (for example, 810nm exposure in preclinical and dermatologic studies), NIR can modulate mitochondrial metabolism, speeding tissue repair and lessening inflammatory responses.

However, it's important to distinguish the therapeutic from the natural context. While forests and green spaces do reflect NIR, the intensity under a canopy is dramatically lower than in medical photobiomodulation.

As a result, any potential effects from outdoor NIR exposure are unlikely to trigger significant biological changes in humans. Any benefits from forest-based NIR would be mild and fleeting, contributing perhaps to a sense of relaxation, rather than delivering the measurable antioxidant or anti-inflammatory outcomes achieved with targeted light therapy.

MEDICINAL FOREST AIR

Forests emit a cocktail of natural chemicals called phytoncides. These include α-pinene, β-pinene, 1,8-cineole and d-limonene, the same aromatic compounds you catch a whiff of when you crush a citrus peel.

Back in the late 2000s, researcher Dr Qing Li and his team in Japan ran a series of 'forest bathing' studies in which people spent several days walking and relaxing among trees. The results were striking: participants' immune system activity kicked up a gear, with natural killer (NK) cells, the body's front-line defenders, becoming more active and more numerous. Levels of key cytotoxic proteins, such as perforin, granulysin and granzymes, also rose. These proteins, used by immune cells to destroy harmful target cells, saw boosts that often lasted a week or more.

Lab tests around the same time hinted at how this might work. Certain essential oils distilled from wood were demonstrated to directly enhance NK cell activity in test-tube experiments, offering a plausible link between breathing in forest air and measurable changes in the immune system. These findings showed that the 'medicine' of the forest isn't just metaphorical – it can be tracked in real biological markers.

AFTER THE RAIN

That vivid smell when rain hits dry ground has a name: petrichor. In 1964, Australian chemists Drs Isabel Bear and Richard Thomas traced the scent to oils in rocks and soils released by rainwater, and soon after to the aromas of chemicals such as geosmin, produced by soil actinomycetes – a group of spore-forming bacteria responsible for that distinctive 'earthy' smell of freshly turned soil. Humans are remarkably sensitive to geosmin. For our prehistoric ancestors, rain following a dry spell might have been the difference between life and death, which could explain why those first drops still evoke such a powerful response today. Do these aromas change health directly? Hard data is thin, because the clinical effects of petrichor remain understudied. Mechanistically, though, gentle, pleasant odours can shape mood and perceived restoration.

TRY THIS:

THE COUNTRYSIDE DOSE GUIDE: A SIMPLE MANUAL FOR DAILY NATURE BENEFITS

Why bother?

Nature isn't just pretty scenery. Spending time among trees, especially those close to home, can quietly boost your health. Research shows:

- **Heart and blood vessel benefits:** greener surroundings encourage walking and moving more, while flora helps to cool the air and clean it.
- **Healthier beginnings:** babies born to mothers living within ~100m of new tree plantings tend to weigh a little more at birth and face lower risks of being too small or arriving too soon.
- **Mind perks:** older, established trees may restore mental calm through richer shade, varied sounds and the general 'settled' feeling they bring.

Administering your countryside dose

- **Access (make it near)**

  The closer the trees, the more they work for you.

  Aim for daily contact – ideally within 100m of your front door or your usual walking route.

- **Density (make it high)**

  Make time to occasionally immerse yourself in woodlands and forests.

  Note: for urban planners the arrangement of trees is key – trees grouped well (lining paths, shading benches, filling small pockets of land) show stronger links to longer life and lower illness rates.

  Go to where there's a mix of heights and species for variety in shade, scent, and wildlife.

- **Maturity (make it old)**

  Older trees typically have larger canopies and greater leaf surface areas, enabling them to intercept and absorb more pollutants, which theoretically increases their pollution removal capacities.

Sensory stacking (your secret boost)

Layering nature's inputs strengthens the effect:

- **See:** green views while walking or from windows.
- **Hear:** birds, leaves in the wind, distant water.
- **Feel:** foliage or cool shade along your route, especially in summer, or hug a tree and feel its texture.
- **Smell:** earthy scents after rain, or freshly mowed grass.

How much and how often?

- **Mini doses:** frequent, short walks in green settings – daily if possible – help mood, focus and attention.
- **Big doses:** a couple of days with multi-hour immersion (think woodland walks or forest stays) can leave short-term imprints on immune and stress markers.

Think of this as a guide, not a prescription. In countryside living, you don't need to measure every step; just keep nature close, varied and regular. Over time, your body and mind will quietly thank you.

WHY ARE YOU SO SENSITIVE?

Hay fever is no small issue. A 2022 global study across 17 countries found that about 14.4 per cent of adults have experienced hay fever at some point in their lives. But that number fluctuates wildly depending on where you live: from just 2.8 per cent in Ibadan, Nigeria, to a whopping 45.7 per cent in Bangkok, Thailand. And the trend is climbing, especially in industrialized nations, with climate change turning what used to be a seasonal annoyance into a growing public health concern.

Imagine not being able to breathe through your nose because of constant stuffiness, relentless sneezing giving you headaches, your eyes so red and itchy that they constantly produce tears, and not because you feel sad. Why would evolution permit such a maladaptive condition to persist in our biology? Could this reaction be a repurposed survival reflex, an ancient system built to detect danger and push it out fast? Hay fever, bothersome as it is, offers a window into that paradox. So, let's see what evolutionary logic there is and what the evidence says about the potential benefits and trade-offs of hay fever.

HERE'S HOW HAY FEVER WORKS

THE FIRST EXPOSURE
When you first encounter pollen (an allergen), certain immune cells (called Th2 cells) sound the alarm. They release chemical signals (cytokines) that tell B cells (antibody factories) to create a very specific weapon: IgE antibodies, custom-built for that pollen (produced by activated B cells, aka plasma cells).

B cells normally produce a range of antibodies, such as IgM (general defence).

THE TRAP IS SET
These IgE antibodies attach to mast cells (tissue-resident immune cells found in your nose, eyes, etc. that store inflammatory signalling compounds like histamine) and basophils (their circulating counterparts in your blood) by docking onto their special FcεRI receptors (the precise 'locks' for the IgE 'key'). Now both these cells become 'sensitized' – like loaded traps waiting for pollen.

THE SECOND EXPOSURE
When pollen hits again, it cross-links with the waiting IgE antibodies (in other words, a single pollen grain is able to physically bridge two adjacent IgE antibodies at once, like two keys turning locks side-by-side). This triggers mast cells to degranulate, meaning they explosively release histamine and other inflammatory mediators.

THE MELTDOWN
Those mediators cause swelling, fluid leaks, itching and sneezing – the classic hay fever symptoms. Talk about overreacting (hypersensitivity) to a harmless visitor.

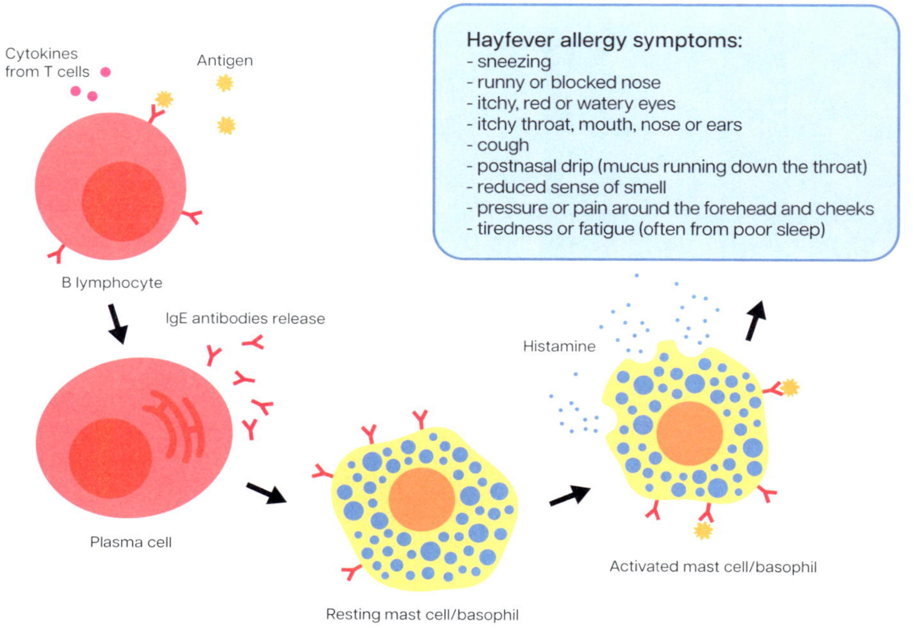

HAY FEVER ALLERGY MECHANISM

The precise definition of what an allergy is continues to be debated. Although IgE-mediated hypersensitivity is central to the definition, leading organizations like the World Allergy Organization (WAO) and the European Academy of Allergy and Clinical Immunology (EAACI) acknowledge that allergy is broader and some allergic diseases involve non-IgE mechanisms. For example, asthma, rhinitis and food allergy vary widely in how they present and what drives them; researchers call these differences 'endotypes'. This heterogeneity in both the symptoms and underlying causes complicates diagnosis and treatment and pushes back against the idea that there is a single mechanism behind all allergies.

- IgE is the linchpin of allergies: no IgE switching ➡ no sensitization of mast cells ➡ no hay fever symptoms.
- IgM is the antibody type normally produced, while B cells only switch to making IgE to defend against parasites (e.g. worms).

A healthy immune system maintains equilibrium between:

- Th1 pathways (fight viruses/bacteria),
- Th2 pathways (fight parasites, drive allergies).

In allergies, this balance is lost ➡ Th2 responses dominate.

Allergies are immune-mediated hypersensitivity reactions to otherwise harmless substances like pollen, dust mites and certain foods. You can think of allergies like a case of mistaken identity in your immune system.

WHY DO ALLERGIES EXIST?

There are two main theories as to why we suffer from allergies. First is the parasite defence hypothesis: IgE and Th2 responses evolved to fight helminths, such as tapeworms, flukes, roundworms and other parasites that once threatened our survival. Second is that they are a defence against toxins, venoms or environmental hazards: quick, unpleasant reactions could help an organism expel or avoid dangerous exposures.

Animal studies suggest allergic responses trigger avoidance behaviour. For example, when an IgE-sensitized mouse develops food aversion, this acts as a built-in 'do not touch' mechanism for risky exposures. This shows that the IgE-mast cell system is built the same way across all mammals, which suggests it served a crucial purpose from the beginning: rapid expulsion of hazards and barrier defence, a biologically sensible adaptation for hostile environments.

These hypotheses suggest that the allergic reactions we experience today are the result of these ancient defence mechanisms within our immune system mistakenly activating when they encounter harmless substances, like pollen.

WHY ARE REACTIONS SO DIFFERENT?

In allergic people, immune cells (like T-helper 2 cells) dominate and release Th2 cytokines (for example, IL-4, IL-5 and IL-13). These cytokines drive allergic inflammation instead of fighting infections. This is why in allergic individuals the immune response is said to skew toward Th2 cytokines. That shift promotes IgE class switching in B cells (they switch from making IgM to IgE) and loads IgE onto FcεRI on mast cells and basophils. Once sensitized, these cells are primed to react. When the allergen appears again, they degranulate rapidly. The mediators they release increase mucus and vascular leakage – fast expulsion defences in action. In the nose, this means congestion, dripping and the urge to sneeze. These are effective if you're trying to flush out a threat. They're somewhat costly if the 'threat' is just a floating cloud of tree pollen.

Allergies are hard to predict, define and treat uniformly because they work differently in different people. Your body has built-in 'peacekeeper' cells (called regulatory T cells or Tregs) that try to maintain balance and prevent overreactions. But allergies are so complex that the peacekeepers must first recognize the unique biological fingerprint of each person's allergy type. This is necessary because not all allergies work in the same way – a concept known as 'endotypic diversity'. For example, one person's reaction might be driven by the classic IgE pathway, while another person's identical symptoms could be caused by a completely different, non-IgE mechanism in the body.

Allergies mean that immune systems react differently, not that a person's immune system is stronger or their defences are better overall. Instead, allergic people's systems are hyper-alert to specific triggers (like pollen) through a process called 'sensitization'. Like a faulty smoke detector blaring at harmless dust, this is a personal misfire, not proof that allergic people have more vigilant immunity in general.

In non-allergic people, Tregs keep Th2 cells in check, which means there is no reaction to pollen.

IN ALLERGIC PEOPLE

- Th2 cells dominate ➡ IgE, mast cell activation ➡ symptoms.
- Tregs are weak/outnumbered ➡ fail to suppress Th2 ➡ allergy persists.

Having an allergy-related biology could offer benefits in certain conditions, but the evidence is not definitive.

ARE THERE ANY ADVANTAGES TO BEING HYPERRESPONSIVE?

There are plausible mechanisms by which allergies could sometimes protect us. One is enhanced immune surveillance, a generally heightened state that might improve detection or elimination of early cancer cells. Another is avoidance learning as discussed earlier (page 179), which reduces risk indirectly.

But scope matters. The benefits of an overactive immune response are highly context dependent and not guaranteed across all situations.

Category	Evidence/findings	Possible mechanisms	Caveats
Infections	Allergic individuals may be less susceptible to parasitic infections (e.g. helminths) due to strong Th2/IgE responses; however, they may be more prone to viral/ bacterial infections due to Th2 immune skewing	• Th2/IgE response effective in expelling parasites	• Protection varies by pathogen type • Immune skewing may impair response to other infections
Cancer	Mixed evidence: some studies show lower risk of certain cancers (due to immune surveillance), others show no change or increased risk	• Enhanced immune surveillance may detect and eliminate early cancer cells	• Results are inconsistent • Context and cancer type matter
Autoimmune diseases	Possibly lower risk of Th1/Th17-driven autoimmunity in those with Th2-dominated allergies; relationship is not consistent	• Immune pathway divergence between allergy (Th2) and autoimmunity (Th1/Th17)	• Not always observed • Disease-specific dynamics

WHY ARE ALLERGIES ON THE RISE?

The hygiene hypothesis posits the idea that, in today's cleaner, more controlled environments, we're not exposed to as many microbes early in life, and this might be one of the reasons why allergies are on the rise. It's thought that there are 'critical windows' in early development, especially during infancy, when our microbiome is taking shape and certain immune pathways are being programmed. What happens in that period could influence how likely we are to develop allergies later on. Large multinational studies (such as GABRIELA and PARSIFAL) suggest something known as the 'farm effect': kids who grow up on farms tend to have a lower risk of hay fever and other allergic conditions. But it's not just farms – having older siblings or living with pets has also been linked to some level of protection. But here's the catch: these effects aren't consistent everywhere, and scientists haven't pinpointed a single microbe or environmental factor that has been consistently verified as protective across different populations and environments.

The exact mechanisms are still not fully understood, and the hygiene hypothesis doesn't explain every pattern, such as why allergy rates are also changing in less affluent areas, or why other immune-related diseases are on the rise. So, while early life microbial exposure does seem to matter, the impact depends heavily on context – including genetics, the type of exposure and when it happens – rather than following one universal rule.

Traits that helped in pathogen- and toxin-rich settings can become mismatched in contemporary environments, where the same hair-trigger reactions misfire as the result of harmless, everyday cues.

HOW TO SHUT YOUR MOUTH FOR 10 DAYS

Going on a silent 10-day meditation retreat might sound like either heaven or hell, depending on your predilection for solitude and exploring your internal state. But thousands of people do it every day in hundreds of devoted centres dotted across the world.

Say an itch shows up on your scalp – an annoying sensation. Your hand immediately wants to scratch it, but, instead, you pause. That itchy feeling intensifies and tests your patience to the very core. After a while, it begins to lessen in its impact, slowly reducing, and then, before you know it, it's gone. Later, a beautiful memory pops up and fades the same way. Moments like this teach something simple and deep: nothing is permanent – everything comes and everything goes. Vipassana, an old, quiet Buddhist practice of paying attention without judgement, uses these tiny lessons through breathwork and interoception to help loosen the pull of craving and aversion, the twin pistons that drive the engine of mental suffering.

ACHIEVING EQUANIMITY

Vipassana is a type of meditation in which you train yourself to notice what's going on, both inside your body and in your mind, in real time. It builds interoceptive awareness (see page 5). This helps you pay attention to sensations, thoughts and feelings as they arise and pass away, without grabbing onto the nice ones (attachment) or pushing away the uncomfortable ones (aversion). A prerequisite to learning the technique as taught by its modern teacher Satya Narayan Goenka is attending a 10-day residential course that provides the essential foundation and deep experience required for proper understanding and practice.

At its core, it's about seeing three basic truths in action:

- Everything changes (impermanence)
- Chasing satisfaction never really works (unsatisfactoriness)
- There's no fixed 'me' running the show (non-self)[9]

Instead of chanting a mantra or locking your focus on just one thing, Vipassana has you stay curious about each moment, watching it arise and fade. Over time, that steady, honest observing helps you step out of old knee-jerk habits and cravings, so you can respond to life with more clarity and less automatic reactivity.

The aim is to achieve equanimity – the ability to still register 'I like this' or 'I don't like that' without feeling pulled around by those reactions and compelled to act on them. By staying balanced in the middle of pleasant and unpleasant experiences, your reactions grow steadier and your perspective widens. This balance not only supports insight and decision making but also fosters greater creativity.

---

[9] The 'you' you feel inside isn't a fixed, permanent entity but a constantly changing flow of thoughts, sensations and experiences that the mind strings together into a story of self.

Scientific research on meditation today tends to focus on what it can measure with instruments, tests or behavioural assessments, rather than on the ultimate spiritual goals described in traditional traditions, such as 'liberation', 'enlightenment' or 'awakening', which refer to states or traits that are inherently subjective and not directly measurable by current science: for example, complete freedom from craving and aversion, insight into the nature of self or abiding equanimity. The ultimate spiritual goals are more valuable than what is scientifically measurable.

THE FOUNDATIONAL ANCHOR – THE BREATH

On a 10-day Vipassana retreat, people start by just noticing their breath – the cool air coming in at the nose, the warm air leaving – or the easy rise and fall of their bellies. They don't have to change anything; they just let it do its thing. The breath becomes a kind of home base, something steady to return to. It helps the mind settle, clears some of the mental chatter and makes you more aware of what's going on inside your body.

In your brain, this gentle attention lights up the parts that handle focus and body awareness, and it loosens the grip of those repetitive thought loops. In Vipassana practice, the breath isn't the destination, it's the entry ramp. Once you're there, you start scanning the whole body, tuning in to every little change in sensation and noticing how those shifts can tug at your thoughts and actions.

IN THE BRAIN

When scientists look at how mindfulness and Vipassana-like practices affect the brain, a few clear patterns emerge:

1 **QUIETER 'STORYTELLING' MODE:** Your brain has a network that's active when you're daydreaming, worrying or thinking about yourself (often called the 'default mode network' or DMN). During meditation, this network dials down. In experienced meditators, it's often quieter even when they're not meditating. This matches reports of less mind-wandering and fewer repetitive thoughts.

2 **STRONGER FOCUS PLUS AWARENESS:** Brain systems involved in paying attention, controlling impulses and tuning into your body sensations appear to get a boost, becoming more active or better connected. Think of this as your brain getting better at staying focused and aware of the present moment.

3 **CALM PLUS STEADY BRAINWAVES:** EEG studies (which measure electrical brain waves) show increased alpha and theta rhythms during meditation. These rhythms are linked to a state of relaxed alertness and steady concentration – exactly what you're aiming for in practice.

A key 2011 study by neuroscientist Dr Judson Brewer of Brown University in the USA really solidified this picture. He found that experienced meditators had clear differences in that 'storytelling' network (DMN), directly linking meditation practice to changes in the brain circuits tied to rumination and mind-wandering. This helped focus research on how meditation might work in the brain.

However, a primary limitation in studies within this field is the reliance on cross-sectional comparisons – research that analyses different groups (like long-term meditators and non-meditators) at a single point in time. This approach makes it challenging to establish causality. More robust, longitudinal research employing well-designed control groups is required to draw definitive conclusions.

STRUCTURAL BRAIN CHANGES

When scientists look at brain structure in long-term meditators, they often find thicker 'focus and feeling' areas. Brain scans show regions linked to attention, emotional balance, body awareness and memory – notably the insula, prefrontal cortex, anterior cingulate and hippocampus – tend to have more grey matter in experienced practitioners. These are the very areas that help us pay attention, manage emotions, sense our bodies and learn. This structural boost offers a plausible biological explanation for the resilience often reported by meditators, and some studies suggest it might even help slow age-related brain decline.

So, how does meditation change your brain? The main theory is that the repeated mental 'workouts' of Vipassana meditation, which involve focusing attention and tuning into body sensations, physically reshape the brain via neuroplasticity. This likely involves strengthening connections between brain cells (synaptic remodelling), building more branches (dendritic growth), improving insulation for faster signals (myelination) and enhancing coordination across brain networks involved in monitoring and control.

Fortunately, it's not just decades-long masters who see positive effects. Studies tracking people over time show that even several months of regular practice can produce measurable shifts in brain structure and how brain networks communicate when simply resting.

> On the mental health side, breath- and body-focused practices are associated with key benefits: reduced stress, anxiety and low mood, as well as improved emotional skills and a healthier sense of self.

But we need to be cautious because it's not possible to say meditation alone causes these brain differences. People drawn to meditation might already have certain brain features, or other healthy lifestyle factors (diet, exercise, sleep) could be playing a role. Many studies have limitations, such as small numbers of participants or difficulty finding perfect comparison groups.

Despite this, these findings are promising, and with more robust research (larger studies and stronger controls) we'll be even better placed to understand whether and how the simple art of observing your inner world as you breathe in and out brings about specific structural changes in your brain. Indeed, that moment where you feel an itch rise and fall points to a bigger skill. One that notices what's here, lets it be and always chooses a response instead of a reflex.

TRY THIS:

ARE YOU READY FOR VIPASSANA?

Question	Yes	Unsure	No	Significance
1. Are you emotionally steady (not in acute grief, trauma or crisis)?	☐	☐	☐	Intensive silence can amplify unresolved emotions – emotional stability helps you cope
2. Are you open to old memories or intense strong feelings surfacing?	☐	☐	☐	Vipassana can unearth stored experiences; acceptance helps you meet them without avoidance
3. Have you used mind-fulness/meditation to handle difficult emotions before?	☐	☐	☐	Prior practice builds skill in staying present when things get turbulent
4. Do you have a support person to talk with afterwards?	☐	☐	☐	Aids integration and smooth return to daily life
5. Can you sit still (~45 min) without much discomfort?	☐	☐	☐	Core part of the practice; readiness limits distraction
6. Do you stretch/move regularly to keep joints flexible?	☐	☐	☐	Prevents stiffness and strain
7. Have you tried multiple 30–45 mins silent sits before?	☐	☐	☐	Reduces body/mind shock on retreat
8. Can you commit to full retreat with no outside contact?	☐	☐	☐	The technique builds progressively; leaving early can disrupt the process
9. Have you scheduled quiet days before and after?	☐	☐	☐	Allows a smoother transition in and out of the retreat environment
10. Have you accepted the retreat's rules (noble silence, early rising, no devices)?	☐	☐	☐	Lowers resistance to structure once you begin

TIP: Treat 'unsure' answers as a sign to explore or prepare further. If you gave many 'no' answers in questions about your emotional stability (Q1–Q4), health (Q5–Q7) or commitment (Q8–Q10), you should consider postponing or starting with smaller goals..

INDEX

KEN OKONA-MENSAH is a freelance science writer and author of *Cracking Anatomy*, who has many years of experience in academia and writing about health-based topics. With a background in pharmacology and toxicology, research and science communication, he previously worked as a scientific writer for Imperial College London where he produced technical and lay reviews for scientific advisory committees of Public Health England. He also sat on the editorial committee for *People & Science* magazine and contributed content relating to public engagement with science and health.

First published in Great Britain in 2026
by Cassell, an imprint of
Octopus Publishing Group Ltd
Carmelite House
50 Victoria Embankment
London EC4Y 0DZ
www.octopusbooks.co.uk

An Hachette UK Company
www.hachette.co.uk

The authorized representative in the EEA is Hachette Ireland, 8 Castlecourt Centre, Dublin 15, D15 XTP3, Ireland (email: info@hbgi.ie)

Distributed in the US by Hachette Book Group 1290 Avenue of the Americas, 4th and 5th Floors, New York, NY 10104

Distributed in Canada by Canadian Manda Group, 664 Annette St., Toronto, Ontario, Canada M6S 2C8

ISBN: 978-1-78840-657-4
eISBN: 978-1-78840-658-1

A CIP catalogue record for this book is available from the British Library.

Printed and bound in China.

10 9 8 7 6 5 4 3 2 1

Commissioned by: Stephanie Selcuk-Frank
Editor: Scarlet Furness
Copy Editor: Paul Murphy
Art Director: Yasia Williams
Designer: Nikki Ellis
Picture Research Manager: Jen Veall
Senior Production Manager: Peter Hunt

Picture Acknowledgements
Alamy Stock Photo: Science History Images 61; iStock: Aldona 96, cnythzl 25, jambojam 10, Liudmila Baskelovich 7, marina_ua 146, NLshop 141, NVS 108, Pikovit44 85, 118, 178, Rujirat Boonyong 26, sv_sunny 87, ttsz 116, 154, VectorMine 22; Shutterstock: Alila Medical Media 46, Andy Frith 112, Blamb 52, KAMAKSI 39, Muhammad Siddique 421 18, Pikovit 94, 137, Piscine26 126, Pranch 11, VectorMine 122.